in Biology,

cluding:

Vindows
r Windows

Statistical and Data Handling Skills in Biology

Third Edition

Roland Ennos

Faculty of Life Sciences, University of Manchester

Harlow, England • London • New York • Boston • San Francisco • Toronto • Sydney • Auckland • Singapore • Hong Kong
Tokyo • Seoul • Taipei • New Delhi • Cape Town • São Paulo • Mexico City • Madrid • Amsterdam • Munich • Paris • Milan

Pearson Education Limited

Edinburgh Gate
Harlow
Essex CM20 2JE
England

and Associated Companies throughout the world

Visit us on the World Wide Web at:
www.pearson.com/uk

First published 2000
Second edition published 2007
Third edition published 2012

ISBN 978-0-273-72949-5

British Library Cataloguing-in-Publication Data
A catalogue record for this book is available from the British Library

Library of Congress Cataloging-in-Publication Data
A catalog record for this book is available from the Library of Congress

10 9 8 7 6 5 4 3 2 1
15 14 13 12 11

Typeset in 9/12.5 by 75
Printed by Ashford Colour Press Ltd., Gosport

Dedication

For my father

Brief contents

Contents

Supporting resources

Visit **www.pearsoned.co.uk/ennos** to find valuable online resources

Companion Website for students

- An Introduction to SPSS version 19 for Windows
- An Introduction to MINITAB version 16 for Windows

For more information please contact your local Pearson Education sales representative or visit **www.pearsoned.co.uk/ennos**

List of figures and tables

Figures

Tables

Preface

It is five years since the second edition of *Statistical and Data Handling Skills in Biology* was first published and I am grateful to Person Education for allowing me the opportunity to update and expand the book for a third edition.

A few more years' experience have prompted me to make some more changes. There were some errors to correct, of course, but the chief failing of the second edition was the artificial separation of parametric and non-parametric tests. In this edition, the book has been restructured to bring the two types of tests together into the same chapters, though in all the cases the parametric tests are introduced first, as this seems logical both from a theoretical and historical perspective. I include more information about the basic examination of distributions, while testing for normality is also brought forward to highlight its importance when deciding which statistical test to perform.

The new edition also includes coverage of additional tests that should take undergraduates up to their final year. There is now coverage of nested ANOVA, the Scheirer–Ray–Hare test, analysis of covariance, and logistic regression, while there is a bigger section on more complex statistical analysis and data exploration. The section on experimental design has also been expanded, with more formal coverage of power analysis.

Finally, there are now comprehensive instructions about how to carry out the statistical tests, not only using the latest version of SPSS (version 19) but also the other common package MINITAB (version 6). I hope that this additional information does not make the book too big or cumbersome.

Like the earlier editions, the book is based on courses I have given to students at the University of Manchester's Faculty of Life Sciences. I am heavily indebted to our e-learning team and to those students who have taken these courses for their feedback. With their help, and with that of several of Pearson Education's reviewers, many errors have been eliminated, and I have learnt much more about statistics, though I take full responsibility for those errors and omissions that remain.

Finally, I would like to thank Yvonne for her unfailing support during the writing of all of the editions of the book.

Standard Deviation, Standard Error, and the **95% Confidence Intervals for Mean.** Note that the package gives some of the items with *too much precision*. Don't copy things from computer screens without thinking!

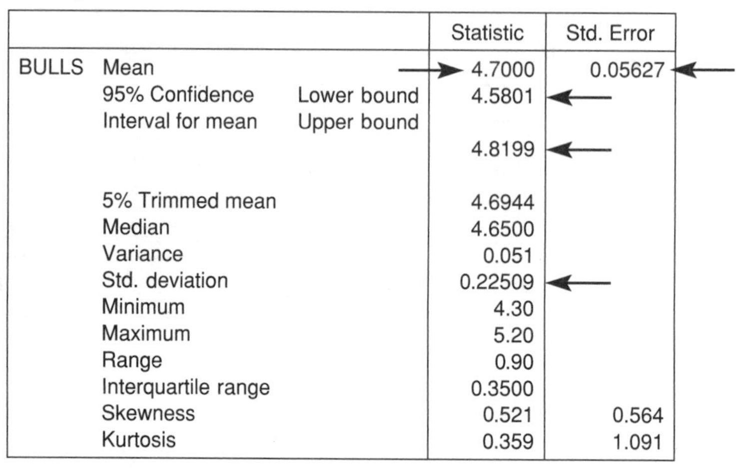

Descriptives

			Statistic	Std. Error
BULLS	Mean		4.7000	0.05627
	95% Confidence Interval for mean	Lower bound	4.5801	
		Upper bound	4.8199	
	5% Trimmed mean		4.6944	
	Median		4.6500	
	Variance		0.051	
	Std. deviation		0.22509	
	Minimum		4.30	
	Maximum		5.20	
	Range		0.90	
	Interquartile range		0.3500	
	Skewness		0.521	0.564
	Kurtosis		0.359	1.091

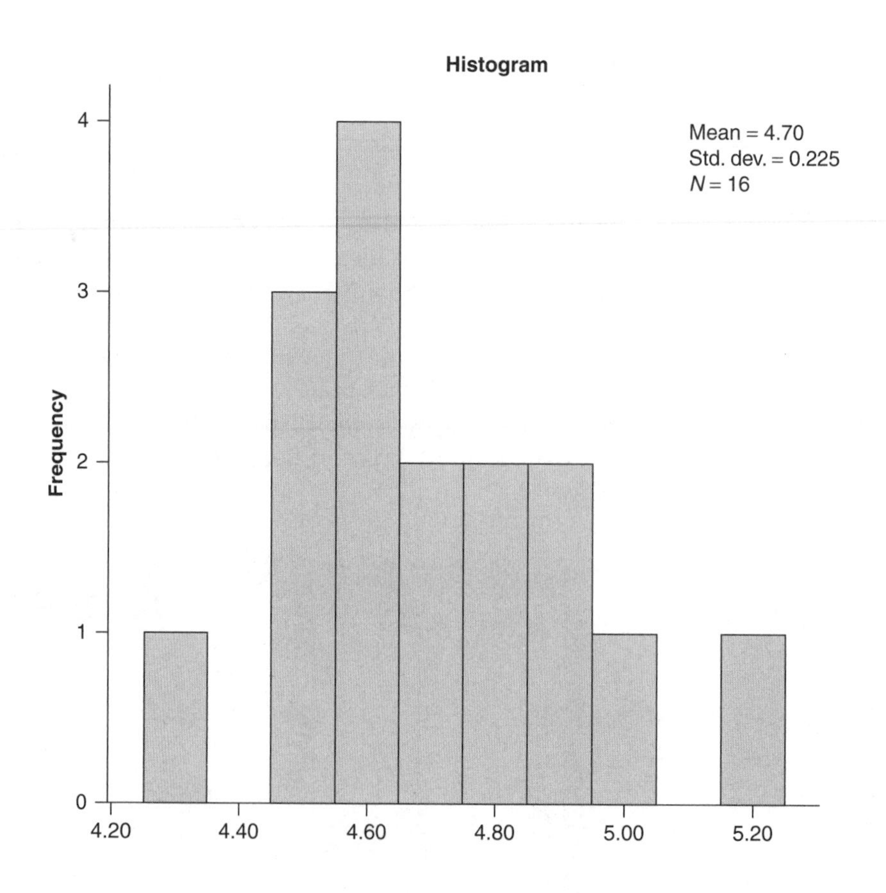

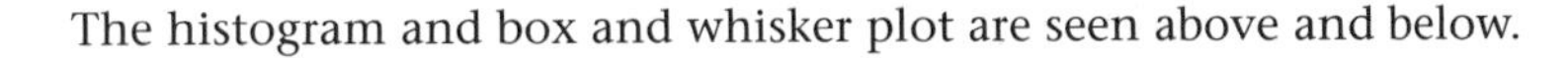

The histogram and box and whisker plot are seen above and below.

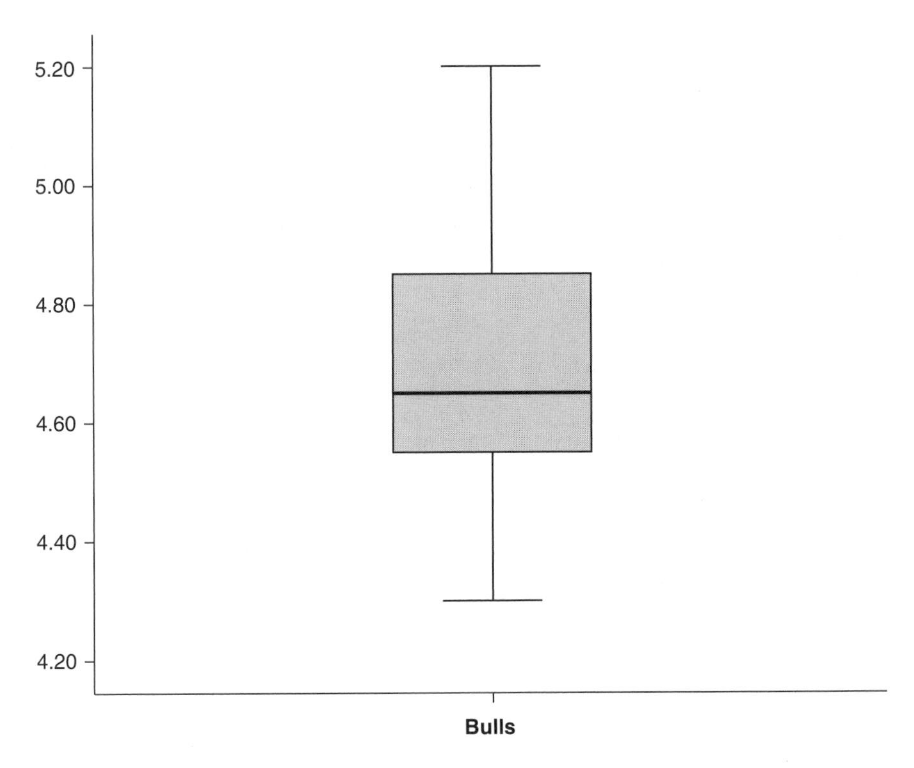

2.9.2 Using MINITAB

Having entered the data into a single column and named it **bulls**, click on the **Stat** menu; from there move onto the **Basic Statistics** bar, then click on **Display Descriptive Statistics.** This brings up the **Display Descriptive Statistics** window.

Click the mouse in the **Variables** box, highlight **C1 bulls** and click on **Select**. This will put **C1 bulls** in the **Variables** box. Next click on the **Graphs** box. This will bring up the **Display Descriptive Statistics – Graphs** window. Tick **Histogram of data** and **Boxplot of data** to give the completed windows below.

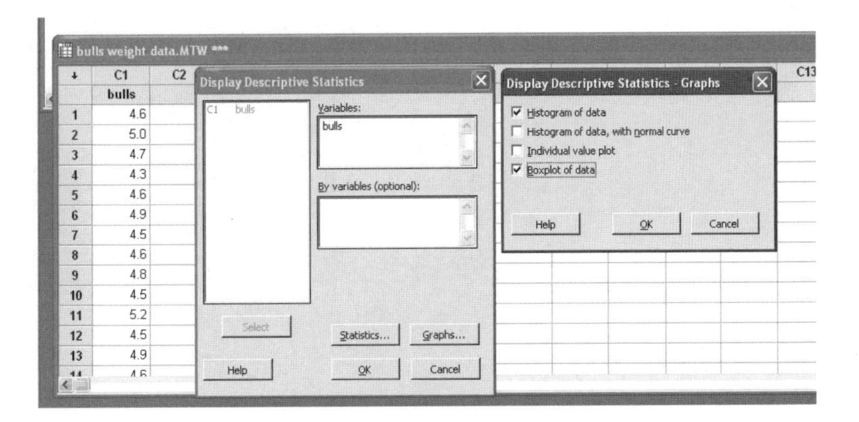

Finally, click on **OK** in the **Display Descriptive Statistics – Graphs** window and **OK** in the **Display Descriptive Statistics** window to perform the calculation.

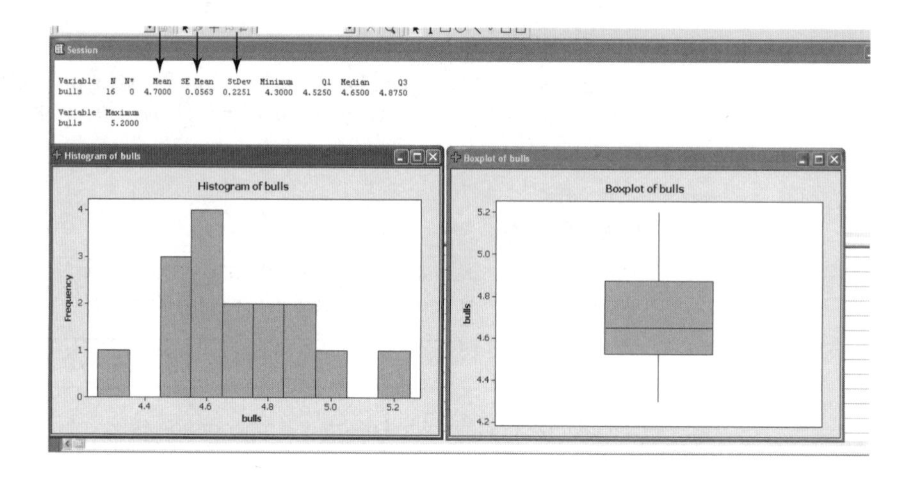

MINITAB will produce the results shown above, giving everything SPSS does except the 95% confidence intervals. If you want those you can go into the **Graphical Summary** dialogue box. MINITAB will then give you the following graphic.

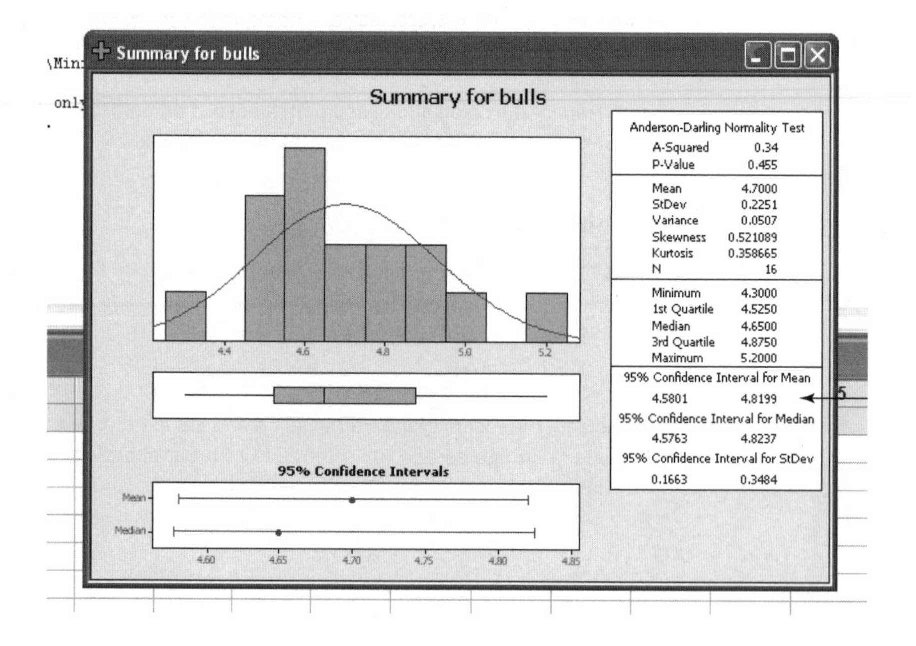

2.10 Self-assessment problems

Problem 2.1

In a population of women, heart rate is normally distributed with a mean of 75 and a standard deviation of 11. Between which limits will 95% of the women have their heart rates?

Problem 2.2

The masses (in grams) for a sample of 10 adult mice from a large laboratory population were measured. The following results were obtained:

5.6	5.2	6.1	5.4	6.3	5.7	5.6	6.0	5.5	5.7

Calculate estimates of the mean and standard deviation of the mass of the mice.

Problem 2.3

Measurements were taken of the pH of nine leaf cells. The results were as follows:

6.5	5.9	5.4	6.0	6.1	5.8	5.8	5.6	5.9

(a) Use the data to calculate estimates of the mean, standard deviation and standard-error of the mean. Use these estimates to calculate the 95% confidence interval for cell pH.
(b) Repeat the calculation assuming that you had only taken the first four measurements. How much wider is the 95% confidence interval?

Problem 2.4

The masses (in kilograms) of 25 newborn babies were as follows.

3.5	2.9	3.4	1.8	4.2	2.6	2.2	2.8	2.9	3.2	2.7	3.4	3.0
3.2	2.8	3.2	3.0	3.5	2.9	2.8	2.5	2.9	3.1	3.3	3.1	

Calculate the mean, standard deviation and standard error of the mean and present your results (a) in figures and (b) in the form of a bar chart with error bars showing standard deviation.

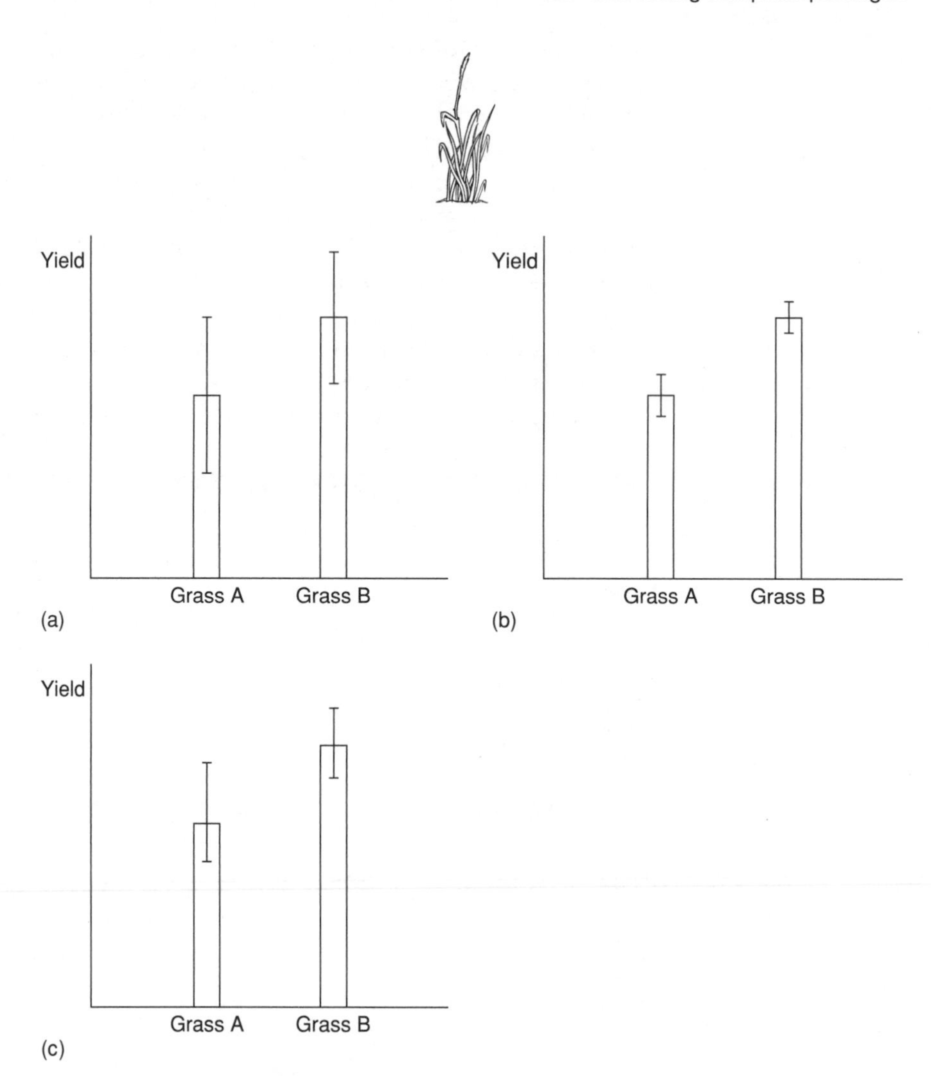

Figure 2.9 Presenting descriptive statistics using error bars. (a) The mean yield of two species of grass with error bars showing their standard deviation; this emphasises the high degree of variability in each grass, and the fact that the distributions overlap a good deal. (b) Standard error bars emphasise whether or not the two means are different; here the error bars do not overlap, suggesting that the means *might* be significantly different. (c) 95% confidence intervals emphasise the likely limits to the mean of each species.

Each package has a different history and its own quirks, so although they have become more similar recently, it is worth introducing them separately, before showing how you can use each of them to calculate descriptive statistics and confidence intervals.

2.8.1 SPSS

SPSS was designed for social scientists and so has a slightly awkward (though very logical) way of working; it assumes you are putting in lots of data about a certain

number of **cases** (usually people). It can then look at how a wide range of factors affect the characteristics of these people (e.g. how does gender affect people's income) and how those characteristics are related to each other (e.g. do richer people tend to be more healthy). Data is entered into a spreadsheet, rather like that of Excel. However, in SPSS, each row represents a particular person (or in biology a particular replicate such as a plant or cell), so you can't put data about two different groups of people or organisms into different columns and use SPSS to analyse it. They have to be identified as members of different groups using another column, and since only numbers are allowed in the spreadsheet, different groups have to be identified by the use of **subscripts**. An example is shown below, giving the heights and genders of eight different people. Here the genders are given as the subscripts 1 and 2 in a separate column, representing female and male.

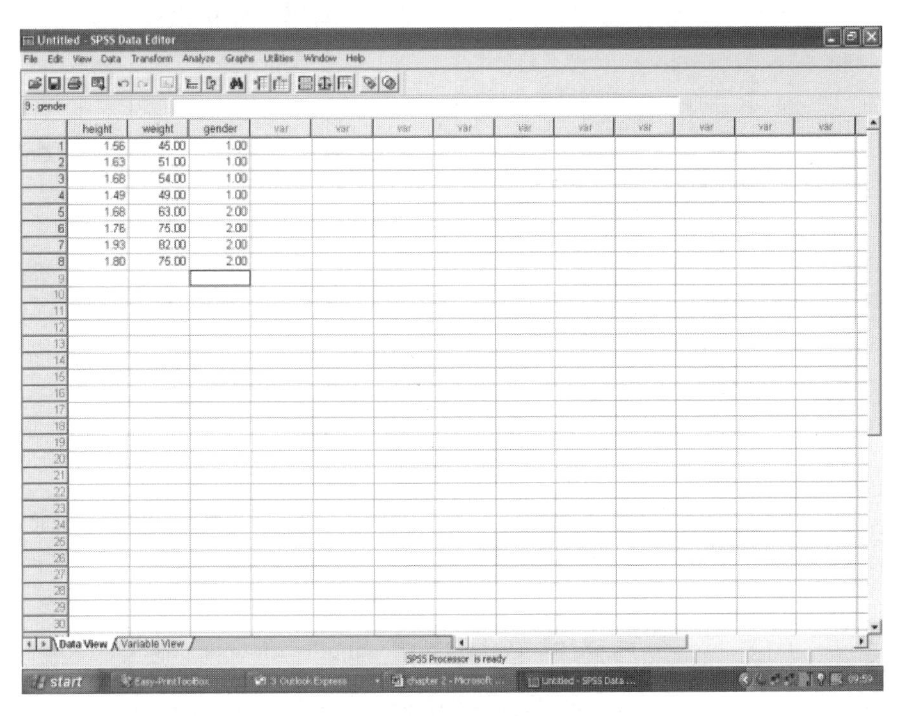

	height	weight	gender
1	1.56	45.00	1.00
2	1.63	51.00	1.00
3	1.68	54.00	1.00
4	1.49	49.00	1.00
5	1.68	63.00	2.00
6	1.76	75.00	2.00
7	1.93	82.00	2.00
8	1.80	75.00	2.00

Source: All SPSS screenshots from SPSS Inc / IBM, Reprint Courtesy of International Business Machines Corporation,© SPSS, Inc., an IBM Company. SPSS was acquired by IBM in October, 2009.

Having entered data into the **Data View Screen** which is shown above (and which appears when you open the program), you name the columns by clicking on the **Variable View** tab to get to the **Variable View Screen** (below) and simply type in the name in the **Name** box. You can also change the width of the column (in the **Width** box) to allow you to get longer numbers on and change the number of decimal places SPSS shows your data to by altering the **Decimals** box.

To return to the data simply click onto the **Data View** tab at the bottom left. You can then order statistical tests to be carried out or plot graphs by pulling down menus with your mouse. The results will be printed out on an **Output Screen**. Both the **Data** and the **Output** can be separately saved.

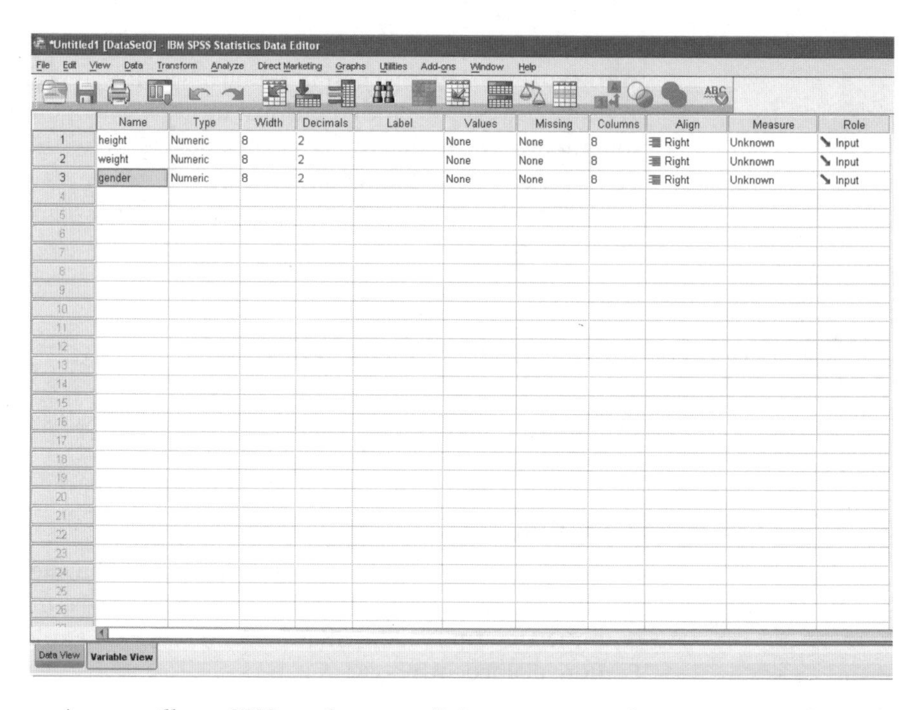

As you will see, SPSS produces prodigious amounts of output, most of it irrelevant to a working biologist. In this book I will present just the useful parts of the results it produces.

2.8.2 MINITAB

MINITAB was designed for scientists and has a slightly different layout from SPSS. You enter your data into a spreadsheet in just the same way, but as it does not expect you to be inputting data about people, it is a bit more flexible (though not always so logical!). You can put data in columns representing different organisms side by side (see below) as well as putting all the data in one column and using subscripts as you do in SPSS.

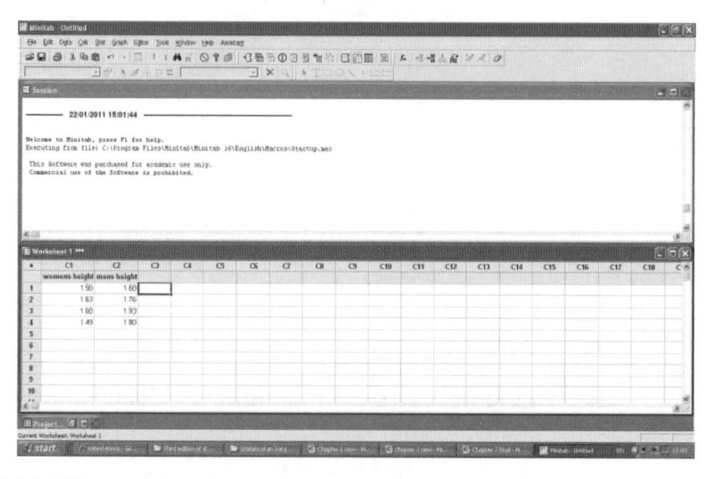

Source: All MINITAB screenshots from MINITAB, portions of the input and output contained in this publication/book are printed with permission of Minitab Inc. All material remains the exclusive property and copyright of Minitab Inc., All rights reserved.

Having entered data into the **Worksheet**, the column titles are simply typed in the boxes below the column number. Finally, you can order statistical tests to be carried out or plot graphs just as in SPSS by pulling down menus with your mouse. The results will be printed out on the **Session** area above the worksheet. Both the **Worksheet** and the **Session** can be saved separately. MINITAB usually produces less output than SPSS, but provides most of the relevant information.

2.9 Calculating descriptive statistics

To demonstrate the workings of the packages we will examine the data on the weight of bull elephants given in Example 2.1.

2.9.1 Using SPSS

Having entered the data into a single column and named it **bulls**, go back into the **Data View Screen**. Next, click on the **Analyze** menu; from there move onto the **Descriptive Statistics** bar, then click on **Explore**. This brings up the **Explore** dialogue window.

Enter the column you want to examine into the **Dependent** box, in this case by clicking on **bulls** in the box on the left and on the arrow pointing to the dependent box. To obtain not only the statistics but also a histogram to show the distribution and a box and whisker plot, click on **Both** in the **Display** box. Then click on the **Plots** box and in the **Explore: Plots** window (below) tick **Histogram** to give the completed windows below. The completed data and windows are shown below.

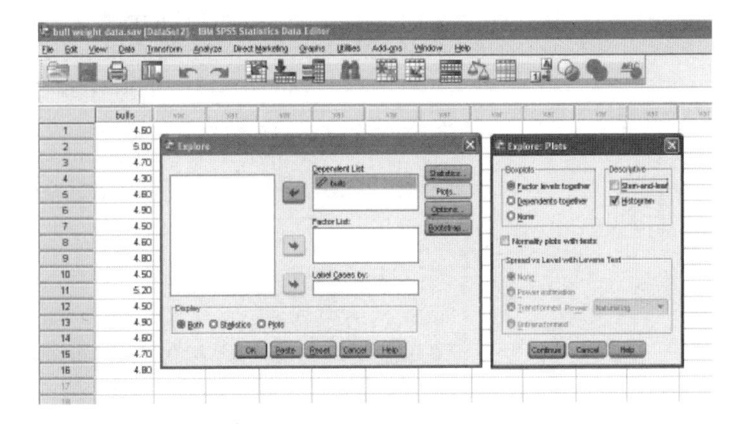

Finally, click on **Continue** in the **Explore: Plots** window and **OK** in the **Explore** window to perform the calculation. SPSS will produce two tables of results, the histogram and the box and whisker plot. Of the tables it is only the second, shown below which is of any use. Take note especially of the **Mean**,

Note how the fluctuations of the cumulative mean start to get less and less and how the line starts to level off. Figure 2.6b shows the cumulative standard deviation. This also tends to level off. If we increased the sample size more and more, we would expect the fluctuations to get less and less until the sample mean converged on the population mean and the sample standard deviation converged on the population standard deviation. The **standard error (SE)** of the mean is a measure of how much the sample means would on average differ from the population mean. Of course, like mean and standard deviation, we cannot know the standard error with any certainty, but we can estimate it. Our **estimate** of the **standard error**, $\overline{SE}$, is given by the expression

standard error (SE)
A measure of the spread of sample means: the amount by which they differ from the true mean. Standard error equals standard deviation divided by the square root of the number in the sample. The estimate of SE is called SE.

$$\overline{SE} = s/\sqrt{N} \tag{2.3}$$

so that the larger the sample size, the smaller the value of as $\overline{SE}$ Figure 2.6c shows. The standard error is an extremely important statistic because it is a measure of just how variable your estimate of the mean is.

2.6 Confidence limits

Once we have our estimate for the mean, $\bar{x}$, and for the standard error, $\overline{SE}$, of the population, it is fairly straightforward to calculate what are known as **confidence limits** for the population mean μ. The most often used are the 95% confidence limits: numbers between which the real population mean, μ will be found 95 times out of 100.

confidence limits
Limits between which estimated parameters have a defined likelihood of occurring. It is common to calculate 95% confidence limits, but 99% and 99.9% confidence limits are also used. The range of values between the upper and lower limits is called the confidence interval.

Because the standard error, $\overline{SE}$, is only estimated, the sample mean will not vary precisely according to the normal distribution, but to a slightly wider one, which is known as the ***t* distribution** (Figure 2.7). The exact shape of the *t* distribution depends on the number of degrees of freedom; it becomes progressively more similar to the normal distribution as the number of degrees of freedom increases (and hence as the estimate of standard deviation becomes more exact).

***t* distribution**
The pattern by which sample means of a normally distributed population tend to vary.

The 95% **confidence limits** for the population mean μ can be found using the tabulated **critical values** of the *t* statistic (Table S1) given in the statistical tables at the end of the book. The critical *t* value $t_{(N-1)}(5\%)$ is the number of standard errors $\overline{SE}$ away from the estimate of population mean $\bar{x}$ within which the real population mean μ will be found 95 times out of 100. The 95% confidence limits define the 95% confidence interval, or 95% CI; this is expressed as follows:

critical values
Tabulated values of test statistics; if the absolute value of a calculated test statistic is usually greater than or equal to the appropriate critical value, the null hypothesis must be rejected.

$$95\%\,\text{CI(mean)} = \text{mean}\ \bar{x} \pm (t_{(N-1)}(5\%) \times \overline{SE}) \tag{2.4}$$

where $(N-1)$ is the number of degrees of freedom. It is most common to use a 95% confidence interval but it is also possible to calculate 99% and 99.9% confidence intervals for the mean by substituting the critical *t* values for 1% and 0.1% respectively into equation 2.4.

Note that the larger the sample size *N*, the narrower the confidence interval. This is because as *N* increases, not only will the standard error $\overline{SE}$ be lower but so will the critical *t* values. Quadrupling the sample size reduces the distance

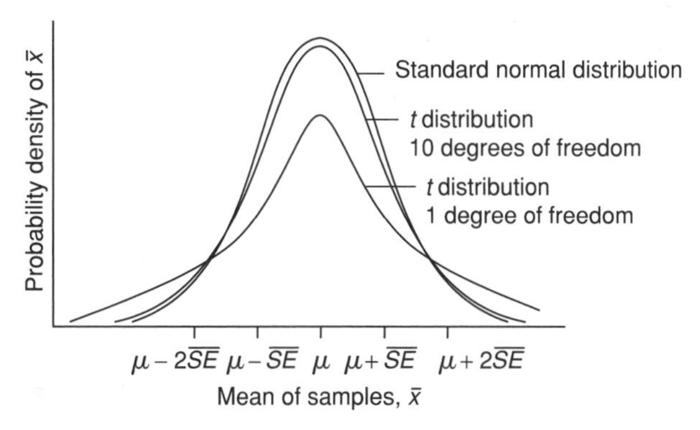

Figure 2.7 Normal distribution and *t* distribution. The distribution of sample means 5 relative to the estimate of the standard error 1 calculated from samples with 1, 10 and infinite degrees of freedom. With infinite degrees of freedom the distribution equals the normal distribution. However, it becomes more spread out as the sample size decreases (fewer degrees of freedom) because the estimate of standard error becomes less reliable.

between the upper and lower limits of the confidence interval by more than one-half.

Take the results for the bull elephants given in Example 2.1. Figure 2.6c shows the standard error of the weights. Note how the standard error falls as the sample size increases. Figure 2.8 shows the cumulative 95% confidence intervals for weight. Note that the distance between the upper and lower intervals falls off extremely rapidly, especially at first; the bigger the sample size the more confident we can be of the population mean.

Example 2.2

Our survey of the 16 bull elephants gave an estimate of mean mass of 4.70 tonnes and an estimate of standard deviation of 0.2251 tonne. We want to calculate 95% and 99% confidence limits for the mean mass.

Solution

The estimate of standard error is $\overline{\text{SE}} = 0.2251/\sqrt{16} = 0.0563$ tonne, which is rounded to 0.056 tonne to three decimal places. Notice that standard errors are usually given to one more decimal place than the mean or standard deviation.

To calculate the 95% confidence limits we must look in Table S1 (p. 258) for the critical value of t for $16-1 = 15$ degrees of freedom. In fact $t_{15}(5\%) = 2.131$. Therefore 95% confidence limits of the population mean are $4.70 \pm (2.131 \times 0.0563) = 4.70 \pm 0.12 = 4.58$ and 4.82 tonnes. So 95 times out of 100 the real population mean would be between 4.58 and 4.82 tonnes.

Similarly, $t_{15}(1\%) = 2.947$. Therefore 99% confidence limits of the population mean are $4.70 \pm (2.947 \times 0.0563) = 4.70 \pm 0.16 = 4.54$ and 4.86 tonnes. So 99 times out of 100 the real population mean would be between 4.54 and 4.86 tonnes.

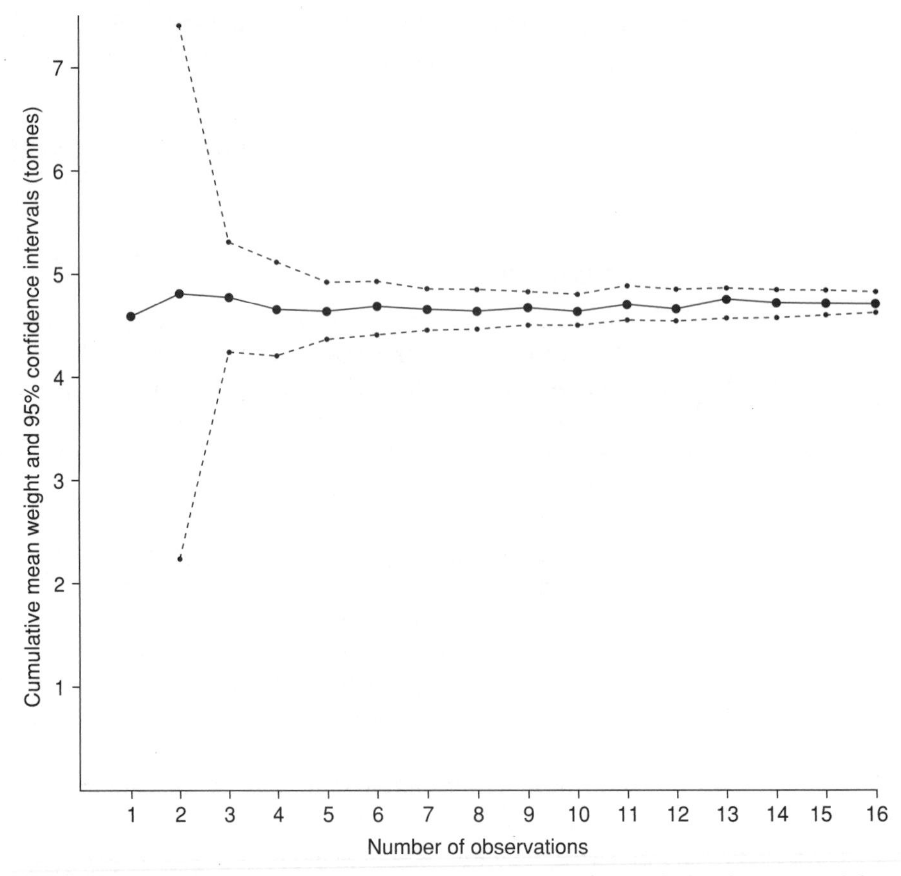

Figure 2.8 Changes in the mean and 95% confidence intervals for the mass of the bull elephants from Example 2.1 after different numbers of observations. Notice how the 95% confidence intervals converge rapidly as sample size increases.

2.7 Presenting descriptive statistics and confidence limits

descriptive statistics Statistics which summarise the distribution of a single set of measurements.

We have seen that it is straightforward to calculate the mean, standard deviation, standard error of the mean (together known as the **descriptive statistics**) and the 95% confidence limits of a sample. Calculating them is the first and most important step in looking at the results of your surveys or experiments. You should work them out as soon as possible and try to see what they tell you.

2.7.1 In text or tables

Once you have obtained your descriptive statistics, you need to express them in the correct way in your write-ups. There are two main ways of doing this. The simplest is just to write them in your text or in tables as the mean followed by the standard deviation or the standard error in parentheses, e.g. $\bar{x}$ (s) or $\bar{x}$ $(\overline{\text{SE}})$. You must say whether you are giving the standard deviation or standard error and you must give the number of observations in your sample; this is so that

the reader can calculate the other statistic. A 95% confidence interval can be given as $\bar{x} \pm (t_{(N-1)}(5\%) \times \overline{SE})$. For example, in our elephant example:

$$\text{Mean and standard deviation} = 4.70\ (0.22)\ t\ (n = 16)$$
$$\text{Mean and standard error} = 4.70\ (0.056)\ t\ (n = 16)$$
$$95\%\ \text{confidence interval} = 4.70\ \pm\ (0.12)\ t\ (n = 16)$$

2.7.2 Graphically

error bars
Bars drawn upwards and downwards from the mean values on graphs; error bars can represent the standard deviation or the standard error.

The other way to present data is on a point graph or a bar chart (Figure 2.9). The mean is the central point of the graph or the top of the bar. **Error bars** are then added. From the mean, bars are drawn both up and down a length equal to either the standard deviation, standard error, or 95% confidence intervals. Finally, lines are drawn across the ends of the bars. You must say in the captions or legends which type of bar you are using.

The choice of which measure of variation to use depends on what you want to emphasise about your results. If you want to show how much **variation** there is, you should choose standard deviation (Figure 2.9a). On the other hand, if you want to show how confident you can be of the mean, you should choose standard error (Figure 2.9b). If you want to show the likely range of the mean you should choose the 95% confidence intervals (Figure 2.9c).

In general, if two samples have overlapping standard error bars, they are unlikely to be statistically different (Chapter 4). Since people in general want to show mean results and tend to want to compare means (see Chapters 4 and 5), standard error bars are by far more the commonly used ones, though some people prefer standard deviation, as it does not hide the variability.

2.8 Introducing computer packages

In the past we used to have to carry out all of our statistical calculations by hand, with the help of statistical tables, and from the 1970s onwards electronic calculators. Fortunately, this is no longer essential (though useful to help you learn about the basics of statistics), because you can use one of the many computer-based statistical packages that are available, such as SPSS (Statistical Package for the Social Sciences), MINITAB, SAS, SYSTAT or R. You simply enter all your results straight into a spreadsheet in the computer package and let the computer take the strain. Using a computer package has two advantages: (1) the computer carries out the calculations more quickly; and (2) you can save the results for future analysis.

In this book we will examine how to carry out statistical tests on two of the most commonly used packages, SPSS and MINITAB. They both work in much the same way. You enter the results from different samples into separate columns of their data spreadsheets. You can then run tests on the different columns from the command screen of the package using a drop-down menu. They run statistical tests, and also produce graphs, which can be useful (though they are rarely of publishable quality).

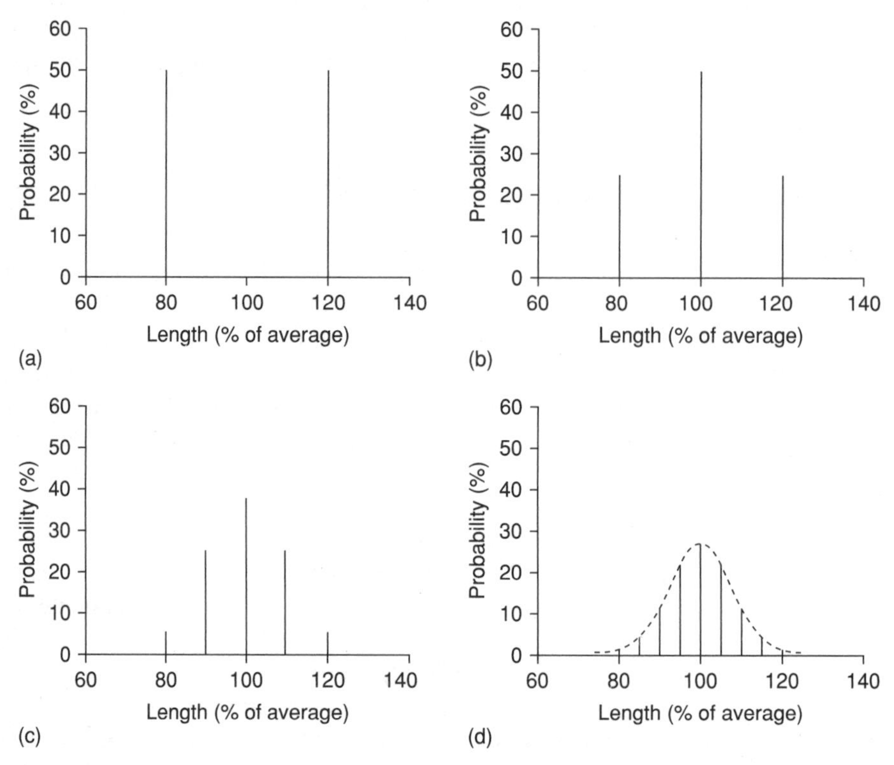

Figure 2.5 Length distributions for a randomly breeding population of rats. Length is controlled by a number of factors, each of which is found 50% of the time in the form which reduces length and 50% in the form which increases length. The graphs show length control by (a) 1 factor, (b) 2 factors, (c) 4 factors and (d) 8 factors. The greater the number of influencing factors, the greater the number of peaks and the more nearly they approximate a smooth curve (dashed outline).

which increases length by 5% and half the time in the form which decreases it by 5%. In this case, of 16 possible combinations of factors, there is only one combination in which all four factors are in the long form and one combination in which all are in the short form. The chances of each are therefore $\frac{1}{2} \times \frac{1}{2} \times \frac{1}{2} \times \frac{1}{2} = \frac{1}{16}$. The rats are much more likely to be intermediate in size, because there are four possible combinations in which three long and one short factor (or three short and one long) can be arranged, and six possible combinations in which two long and two short factors can be arranged. It can be seen that the central peak is higher than those further out. The process is even more apparent, and the shape of the distribution becomes more obviously humped if there are eight factors, each of which increases or decreases length by 2.5% (Figure 2.5d). The resulting distributions are known as **binomial distributions**.

binomial distributions The pattern by which the sample frequencies in two groups tends to vary.

If length were affected by more and more factors, this process would continue; the curve would become smoother and smoother until, if length were affected by an infinite number of factors, we would get the bowler-hat-shaped distribution curve we saw in Figure 2.4. This is the so-called **normal distribution** (also known as the *Z* distribution). If we measured an infinite number of rats, most would have length somewhere near the average, and the numbers would tail off on each

side. Because so many biological characteristics are influenced by large numbers of genes and environmental factors, many follow the normal distribution more or less closely.

2.4 Describing the normal distribution

parameters
A measure, such as the mean and standard deviation, which describes or characterises a population. These are usually represented by Greek letters.

population
A potentially infinite group on which measurements could be taken. Parameters of populations usually have to be estimated from the results of samples.

sample
A subset of a possible population on which measurements are taken. These can be used to estimate parameters of the population.

estimate
A parameter of a population which has been calculated from the results of a sample.

statistics
An estimate of a population parameter, found by random sampling. Statistics are represented by Latin letters.

variance
A measure of the variability of data: the square of their standard deviation.

degrees of freedom (DF)
A concept used in parametric statistics, based on the amount of information you have when you examine samples. The number of degrees of freedom is generally the total number of observations you make minus the number of parameters you estimate from the samples.

Unlike general distributions which need at least five numbers to describe them, the normal distribution of a population can be described by just two numbers or **parameters**. The position of the centre of the distribution is described by the **population mean** μ, which on the graph is located at the central peak of the distribution (Figure 2.4). The width of the distribution is described by the **population standard deviation** σ, which is the distance from the central peak to the point of inflexion of the curve (where it changes from being convex to concave) (Figure 2.4). This is a measure of about how much, on average, points differ from the mean. Of course we can never know for certain the population parameters because we would never have the time to measure the entire population, but we can use the results from a **sample** of a manageable size to make an **estimate** of the population mean and standard deviation. These estimates are known as **statistics**.

It is very easy to calculate an **estimate of the population mean**. It is simply the average of the sample, or the sample mean $\bar{x}$. It is simply the sum of all the lengths divided by the number of rats measured. In mathematical terms this is given by the expression

$$\bar{x} = \frac{\sum x_i}{N} \tag{2.1}$$

where x_i is the values of length and N is the number of rats.

The **estimate of the population standard deviation**, written s or σ_{n-1}, is given by the expression

$$s = \sigma_{n-1} = \sqrt{\frac{\sum (x_l - \bar{x})^2}{N - 1}} \tag{2.2}$$

It is the square root of the **variance**, which is the average of the square of the distances of each point from the sample mean. Rather than dividing the sum of squares by N, however, we divide by $(N-1)$. We use $(N-1)$ because this expression will give an unbiased estimate of the population standard deviation, whereas using N would tend to underestimate it. To see why this is so, it is perhaps best to consider the case when we have only taken one measurement. Since the estimated mean $\bar{x}$ necessarily equals the single measurement, the standard deviation we calculate when we use N will be zero. Similarly, if there are two points, the estimated mean will be constrained to be exactly halfway between them, whereas the real mean is probably not. Thus the variance (calculated from the square of the distance of each point to the mean) and hence standard deviation will probably be underestimated.

The quantity $(N-1)$ is known as the number of **degrees of freedom** of the sample. Since the concept of degrees of freedom is repeated throughout the rest of this book, it is important to describe what it means. In a sample of N obser-

vations each is free to have any value. However, if we have used the measurements to calculate the sample mean, this restricts the value the last point can have. Take a sample of two measurements, for instance. If the mean is 17 and the first measurement is 17 + 3 = 20, the other measurement *must* have the value 17 − 3 = 14. Thus, knowing the first measurement fixes the second, and there will only be one degree of freedom. In the same way, if you calculate the mean of any sample of size N, you restrict the value of the last measurement, so there will be only $(N-1)$ degrees of freedom.

It can take time calculating the standard deviation by hand, but fortunately few people have to bother nowadays; estimates for the mean and standard deviation of the population can readily be found using computer statistics packages or even scientific calculators. All you need to do is type in the data values and press the $\bar{x}$ button for the mean and the s, σ_{n-1} or $x_{\sigma n-1}$ button for the population standard deviation. Do not use the σ_n or $x_{\sigma n}$ button, since this works out the sample standard deviation, NOT the population standard deviation.

Example 2.1

The masses (in tonnes) of a sample of 16 bull elephants from a single reserve in Africa were as follows.

4.6	5.0	4.7	4.3	4.6	4.9	4.5	4.6
4.8	4.5	5.2	4.5	4.9	4.6	4.7	4.8

Using a calculator, estimate the population mean and standard deviation.

Solution

The estimate for the population mean is 4.70 tonnes and the population standard deviation is 0.2251 tonne, rounded to 0.23 tonne to two decimal places. Note that both figures are given to one more degree of precision than the original data points because so many figures have been combined.

2.5 The variability of samples

It is relatively easy to calculate estimates of a population mean and standard deviation from a sample. Unfortunately, though, the estimate we calculated of the population mean $\bar{x}$ is unlikely to exactly equal the real mean of the population. In our elephant survey we might by chance have included more light elephants in our sample than one might expect, or more heavy ones. The estimate itself will be variable, just like the population. However, as the sample size increases, the small values and large values will tend to cancel themselves out more and more. The estimated mean will tend to get closer and closer to the real population mean (and the estimated standard deviation will get closer and closer to the population standard deviation). Take the results for the bull elephants given in Example 2.1. Figure 2.6a shows the cumulative mean of the weights.

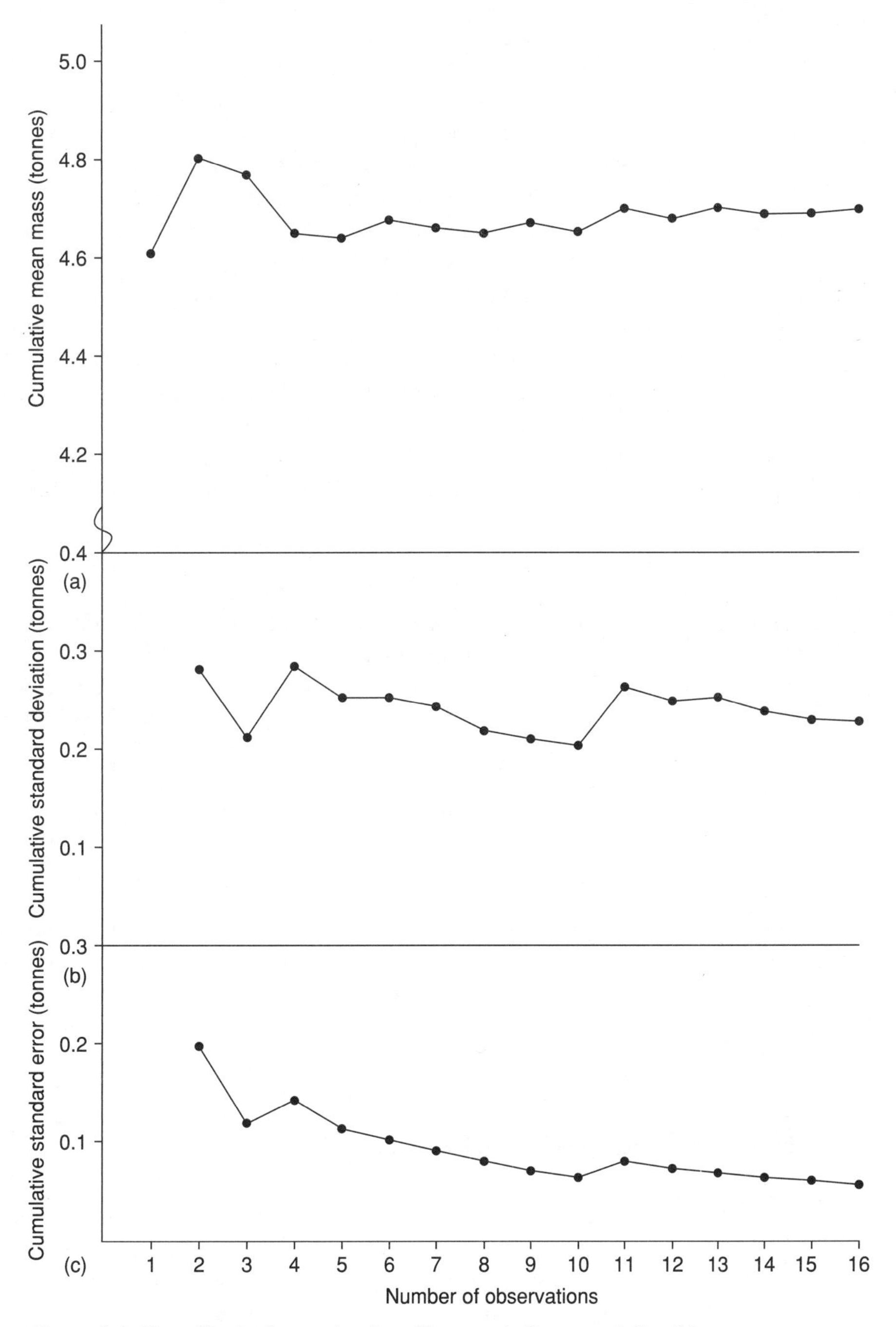

Figure 2.6 The effect of sample size. Changes in the cumulative (a) mean, (b) standard deviation and (c) standard error of the mass of bull elephants from Example 2.1 after different numbers of observations. Notice how the values for mean and standard deviation start to level off as the sample size increases, as you get better and better estimates of the population parameters. Consequently the standard error (c), a measure of the variability of the mean, falls.

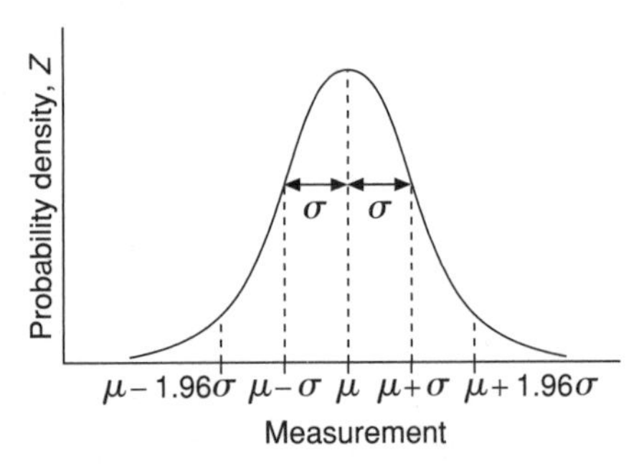

Figure 2.4 A normal distribution. The centre of the distribution is described by the mean μ and the width by the standard deviation σ which is the distance to the point of inflexion: 68% of measurements are found within one standard deviation of the mean; 95% are found within 1.96 times the standard deviation of the mean.

looking at just why data is normally distributed, and then examining how normally distributed data is described.

2.3.1 Why characteristics are normally distributed

There are three main reasons why the measurements we take of biological phenomena vary. The first is that organisms differ because their genetic make-up varies. Most of the continuous characters, like height, weight, metabolic rate or blood [Na^+], are influenced by a large number of genes, each of which has a small effect; they act to either increase or decrease the value of the character by a small amount. Second, organisms also vary because they are influenced by a large number of environmental factors, each of which has similarly small effects. Third, we may make a number of small errors in our actual measurements.

distribution
The pattern by which a measurement or frequency varies.

So how will these factors influence the **distribution** of the measurements we take? Let's look first at the simplest possible system; imagine a population of rats whose length is influenced by a single factor that is found in two forms. Half the time it is found in the form which increases length by 20% and half the time in the form which decreases it by 20%. The distribution of heights will be that shown in Figure 2.5a. Half the rats will be 80% of the average length and half 120% of the average length.

What about the slightly more complex case in which length is influenced by two factors, each of which is found half the time in a form which increases length by 10% and half the time in a form which decreases it by 10%? Of the four possible combinations of factors, there is one in which both factors increase length (and hence length will be 120% of average), and one in which they both reduce length (making length 80% of average). The chances of being either long or short are $\frac{1}{2} \times \frac{1}{2} = \frac{1}{4}$. However, there are two possible cases in which overall length is average: if the first factor increases length and the second decreases it; and if the first factor decreases length and the second increases it. Therefore 50% of the rats will have average length (Figure 2.5b).

Figure 2.5c gives the results for the even more complex case when length is influenced by four factors, each of which is found half the time in the form

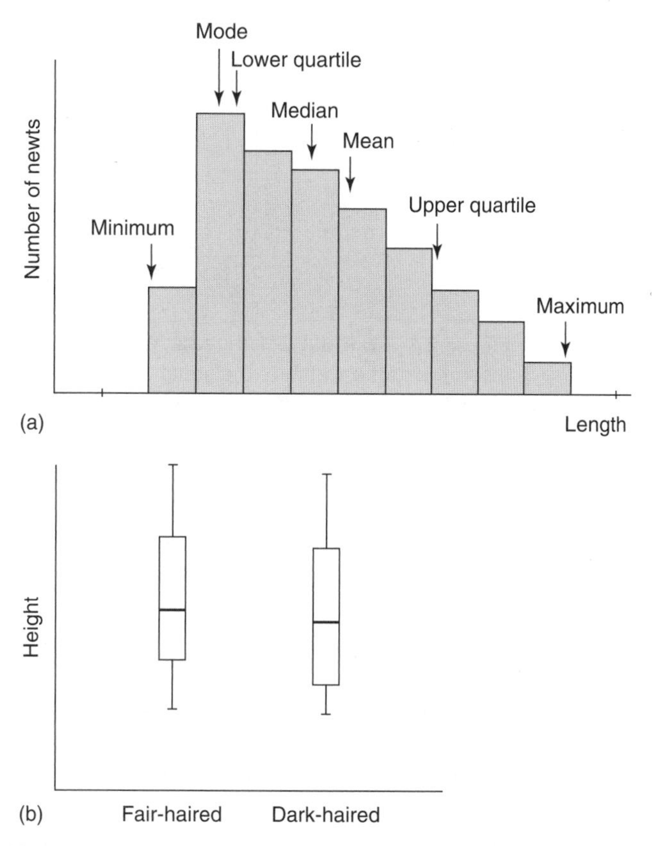

Figure 2.3 Measurements of the distribution of data. The median, quartiles and maximum and minimum values of the positively skewed distribution (a) are best summarised using a box and whisker plot (b), such as this which compares the height of fair-haired and dark-haired women.

to see how symmetrical the distribution is, and how much the data is concentrated towards the middle. Giving two box and whisker plots side by side of two different samples also allows you to compare them at a glance. In Figure 2.3b, for instance, it is clear that there is not really that much difference between fair-haired and dark-haired women.

2.3 The normal distribution

normal distribution The usual symmetrical and bell-shaped distribution pattern for measurements that are influenced by large numbers of factors.

When biologists first seriously started to investigate variability at the end of the nineteenth century, they quickly discovered that a great number of characteristics of organisms varied according to the **normal distribution**. This is a symmetrical, bowler hat-shaped distribution (Figure 2.4) with the numbers falling off in a bell curve either side of the mean.

Because the normal distribution is so important, and so many statistical tests assume that data is normally distributed, I think it is worth spending some time

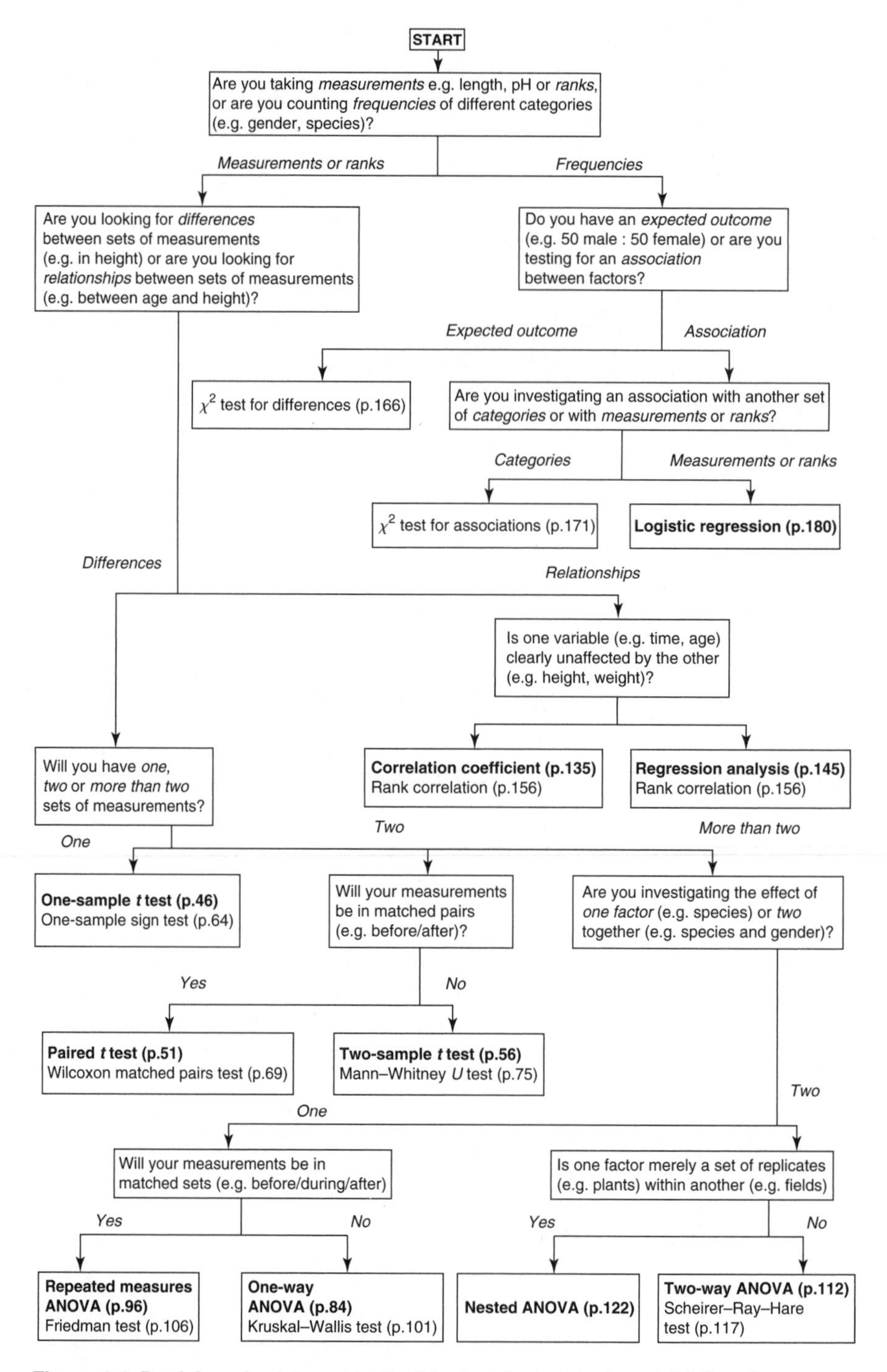

Figure 1.1 Decision chart for statistical tests. Start at the top and follow the questions down until you reach the appropriate box. The tests in normal type are non-parametric equivalents for irregularly distributed or ranked data.

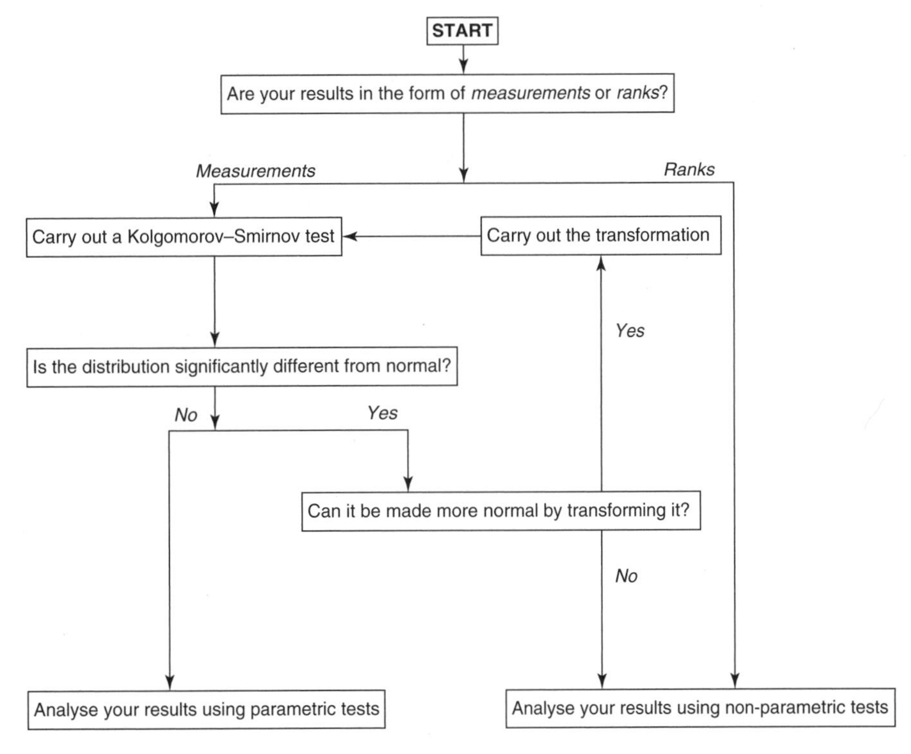

Figure 1.2 Flow chart showing how to deal with measurements and rank data. Start at the top, answer the questions and transform data where appropriate before deciding whether you can use parametric tests or have to make do with non-parametric ones.

be transformed to the normal distribution, or have measurements which can only have a few, discrete, values. Before finally deciding which tests to carry out, therefore you need to investigate the distribution of your data (Figure 1.2 and on the inside cover of the book) and see whether it is valid to carry out parametric tests, or if it is possible to transform your data so that you can.

1.8 Using this book

1.8.1 Carrying out tests

Once you have made your decision, the chart will direct you to a page in the main section of this book (Chapters 4–7), which describes the main statistical tests. You should go to the page indicated, where details of the test will be described. Each test description will do five things.

1. It will tell you the sorts of questions the test will enable you to answer and give examples to show the range of situations for which it is suitable. This will help you make sure you have chosen the right test.
2. It will tell you when it is valid to use the test.

3. It will describe the rationale and mathematical basis for the test; basically it will tell you how it works.
4. It will show you how to perform the test using a calculator and/or the computer-based statistical packages SPSS and MINITAB.
5. It will tell you how to present the results of the statistical tests.

1.8.2 Designing experiments

As a research biologist you will not only have to choose statistical tests and perform the analysis yourself; you will also have to design your own experiments. Chapter 8 will show how you can use the information about statistics set out in the main part of the book to design better experiments.

1.8.3 Complex statistical analysis

This book describes most of the statistical tests you will need to analyse straightforward experiments and surveys: ones that look at one or at most two factors. I would strongly recommend that you stay as far as possible within these limits. There may be some occasions, however, particularly within some branches of biology, where you simply have to carry out and analyse rather more complex experiments or have to investigate huge sets of data. Chapter 9 describes some of the complex statistical techniques that can help you investigate several factors at once.

1.8.4 Manipulating numbers and units

Chapter 10 describes how you should manipulate numbers and units, a skill which is often a prerequisite to dealing with data, even before you can think about statistical analysis.

Before you can carry out statistical tests, however, you need to know how to deal with and quantify variability, and to investigate how and why organisms vary in the first place. This is all set out in Chapter 2.

2 Dealing with variability

2.1 Introduction

This chapter tells you how to deal with the problem of variability: it shows how to examine and present the distribution of data; explains why variation occurs in the first place; and describes how to quantify it. In other words, it shows how you can obtain useful quantitative information about a population from the results of your sample, despite the variation.

2.2 Examining the distribution of data

The first thing to do when you have taken some measurements from your sample is to investigate their distribution. The best way of doing this is produce a **histogram** or **bar chart**.

For continuous data, you should produce a histogram (Figure 2.1a), grouping the data points into a number of arbitrarily defined size classes of equal width set out along the x-axis, while the y-axis shows the number of data points in each class. This gives very useful information about the distribution, in particular about the relative commonness of different values. The number of classes you choose should depend on the sample size. If you have a very large sample you could have anything up to 12 or more classes to produce a detailed distribution. However, with smaller sample sizes the numbers within each class fall and the distribution is likely to become more bumpy. It is better, then, to have a smaller number of classes: as few as 5 for small samples of 20 or less.

Discrete data can be treated in just the same way as continuous data, with each class covering the same number of discrete values. However, if you have a big enough sample, each discrete value may have enough data points within it to allow you to draw a bar chart (Figure 2.1b), in which each bar is separated from the next.

2.2.1 Types of distribution

The next step is to examine the distribution that your histogram reveals. There are many ways in which your data could be distributed. It could be **symmetrically**

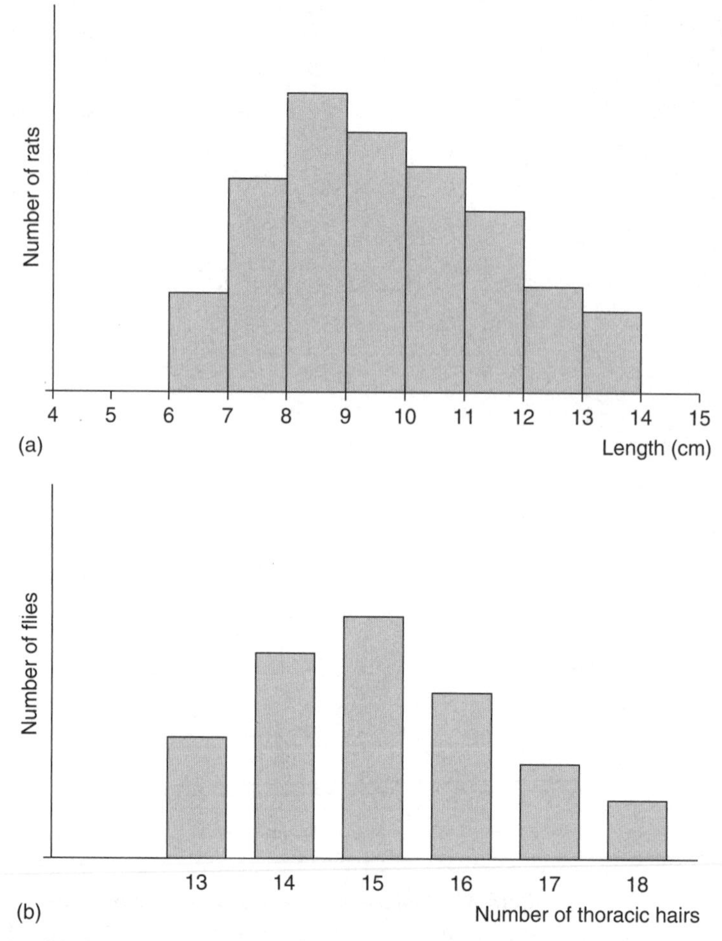

Figure 2.1 Methods of presenting the distribution of a sample. Continuous data should be presented as a histogram (a) which gives the numbers of points within a number of classes of equal width. Discrete data can instead be given in a bar chart (b).

distributed (Figure 2.2a), **positively skewed**, with more small values than large ones. (Figure 2.2b), or **negatively skewed**, with more large values than small ones (Figure 2.2c). Positively skewed data are particularly commonly in natural populations because there tend to be many more young (and hence small) organisms in a population than older, larger ones. Data may even be **irregularly** distributed (Figure 2.2d).

Whichever way the data is distributed, there is no way that anyone else would be particularly interested in seeing all your histograms; you need a way to summarise and quantify the distribution.

2.2.2 Representing distributions

The *width* of the distribution can be summarised by giving the **maximum** and **minimum** values (Figure 2.3a). The *centre* of the distribution is more tricky to represent, and there are three different values that you could give. The **mode** is

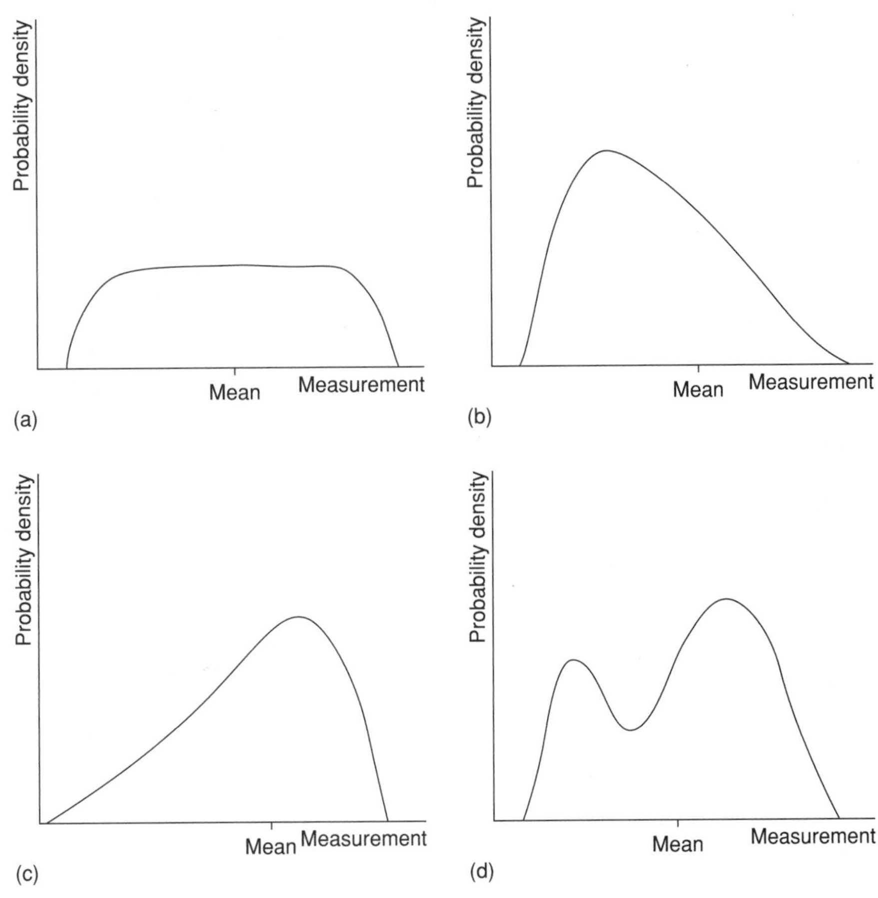

Figure 2.2 Different ways in which data may be distributed. (a) A symmetrical distribution; (b) positively skewed data; (c) negatively skewed data; (d) irregularly distributed data.

mean (μ)
The average of a population. The estimate of μ is called $\bar{x}$.

skewed data
Data with an asymmetric distribution.

median
The central value of a distribution (or average of the middle points if the sample size is even).

quartiles
Upper and lower quartiles are values exceeded by 25% and 75% of the data points, respectively.

the class in which there are the most data points. I don't recommend you use the mode, as its value will depend on exactly how you have split up your data into size classes. The **mean** is the arithmetical average of all the data points. As we shall see, in many cases this is extremely useful, but it is not very helpful for **skewed data**, when the mean will be greatly affected by the few outlying points. The most universally useful measure of the centre of the distribution is the **median** which is the point halfway along the ranked data set (or the average of the points above and below the middle if the sample size is even). Finally, the *shape* of the distribution is best represented by finding the **quartiles**, the points 25% and 75% down the ranked data set. The **interquartile range** is the distance between these two points, and is another measure of the *width* of the distribution.

These measures can be combined to produce a box and whisker plot (Figure 2.3b) with the median as a thick bar at the centre, the upper and lower quartiles as the top and bottom of the box, and the maximum and minimum values as the top and bottom of the whiskers. This one simple plot allows you

controls. Or you might want to compare two sets of measurements taken on a single group, for instance before and after subjecting people to some medical treatment. Tests to answer these questions are described in Chapter 4. Alternatively, you might want to see if organisms of several different types (for instance five different bacterial strains) or ones that have been subjected to several types of treatments (for instance wheat subjected to different levels of nitrate and phosphate) are different from each other. Tests to answer these questions are described in Chapter 5.

(b) Testing for relationships between measurements Another thing you might want to do is to take two or more measurements on a single group of organisms or cells and investigate how the measurements are **related**. For instance, you might want to investigate how people's heart rates vary with their blood pressure; how weight varies with age; or how the concentrations of different cations in neurons vary with each other. This sort of knowledge can help you work out how organisms operate, or enable you to predict things about them. Chapter 6 describes how statistical tests can be used to quantify relationships between measurements and work out if the apparent relationships are real.

(c) Testing for differences and relationships between categorical data There are three different things you might want to find out about categorical data. You might want to determine whether there are different frequencies of organisms in different categories from what you would expect; do rats turn more frequently to the right in a maze than to the left, for instance. Alternatively you might want find out whether categorical traits, for instance people's eye and hair colour, are associated: are people with dark hair also more likely to have brown eyes? Finally, you might be interested in working out how quantitative measurements might affect categorical traits, for instance are tall people more likely to have brown eyes? Tests to answer all these sorts of questions are described in Chapter 7.

1.7 Using the decision chart

The logic of the previous section has been developed and expanded to produce a **decision chart** (Figure 1.1 and on the inside cover of the book). Though not fully comprehensive, the chart includes virtually all of the tests that you are likely to encounter as an undergraduate. If you are already a research biologist, it may also include all the tests you are ever likely to use over your working life! If you follow down from the start at the top and answer each of the questions in turn, this should lead you to the statistical test you need to perform.

There is only one complication. The final box may have two alternative tests: a parametric test in bold type and an equivalent non-parametric test in normal type. You are always advised to use the parametric test if it is valid, because parametric tests are more powerful in detecting significant effects. Use the non-parametric test if you are dealing with ranked data, irregularly distributed data that cannot

statistics books such as this one contain large numbers of different tests. Why are there so many? There are two main reasons for this. First, there are several very different ways of quantifying things and hence different types of data that you can collect, and this data can vary in different ways. Second, there are very different questions you might want to ask about the data you have collected.

1.6.1 Types of data

(a) Measurements The most common way of quantifying things about organisms is to take **measurements** (of things such as height, mass or pH), to give what is also known as **interval data**. This sort of data can vary **continuously**, like weight (e.g. 21.23 or 34.651 kg) or **discretely**, like the numbers of hairs on a fruit fly (e.g. 12 or 18). As we shall see in Chapter 2, many of these measurements vary according to the **normal distribution**. There is a set of tests, the so-called **parametric tests**, that assume that this is the case. On the other hand, many measurements do not vary in this way. This sort of data either has to be **transformed** until it does vary according to the normal distribution (Chapter 3) or, if that is not possible, it must be analysed using a separate set of tests, the **non-parametric tests**, which make no assumption of normality.

measurements
A character state which can meaningfully be represented by a number.

normal distribution
The usual symmetrical and bell-shaped distribution pattern for measurements that are influenced by large numbers of factors.

parametric tests
A statistical test which assumes that data are normally distributed.

non-parametric tests
A statistical test which does not assume that data is normally distributed, but instead uses the ranks of the observations.

(b) Ranks On many occasions, you may only be able to put your measurements into an order, without the actual values having any real meaning. This **ranked** or **ordinal data** includes things like the pecking order of hens (e.g. 1st, 12th), the seriousness of an infection (e.g. none, light, medium, heavy) or the results of questionnaire data (e.g. 1 = poor to 5 = excellent). This sort of data *must* be analysed using **non-parametric tests.**

(c) Categorical data Some features of organisms are impossible to quantify in any way. You might only be able to classify them into different **categories**. For instance birds belong to different species and have different colours; people could be diseased or well; and cells could be mutant or non-mutant. The only way of quantifying this sort of data is to count the ***frequency*** with which each category occurs. This sort of data is usually analysed using χ^2 (chi-squared) tests or logistic regression (Chapter 7).

frequency
The number of times a particular character state turns up.

1.6.2 Types of questions

There are two main types of questions that statistical tests are designed to answer. Are there differences between sets of measurement? or are there relationships between them?

(a) Testing for differences between sets of measurements There are many occasions when you might want to test to see whether there are **differences** between two groups, or types of organisms. For instance, we have already looked at the case of comparing the height of fair- and dark-haired women. An even more common situation is when you carry out experiments; you commonly want to know if experimentally treated organisms or cells are different from

Publisher's acknowledgements

We are grateful to the following for permission to reproduce copyright material (t = top, c = centre, b = bottom):

SPSS screenshots on pages 24, 25 (t), 32, 37 (t, b), 39 (t, b), 48, 53, 59 (t, b), 66, 71, 76, 88, 93, 97, 98, 103, 108, 114, 119, 124 (t, b), 139, 148, 158, 168 (t, b), 173, 174 (t, c, b), 181, 206, 209 from SPSS Inc / IBM, Reprint Courtesy of International Business Machines Corporation,© SPSS, Inc., an IBM Company. SPSS was acquired by IBM in October, 2009.

MINITAB screenshots on pages 25 (b), 26, 28, 29 (t, b), 33 (t, b), 36 (t, b), 40 (t, b), 41 (t, b), 49, 54, 61, 67, 72, 73, 77, 89, 94, 104, 110, 115, 120, 126, 140, 141, 142, 143, 149, 159 (t, b), 169, 175, 183, 196, 207 from MINITAB, portions of the input and output contained in this publication/book are printed with permission of Minitab Inc. All material remains the exclusive property and copyright of Minitab Inc., All rights reserved.

In some instances we have been unable to trace the owners of copyright material and we would appreciate any information that would enable us to do so.

1 An introduction to statistics

1.1 Becoming a research biologist

A biologist can be defined as someone who studies the living world. Much of a biologist's training involves learning about what other people have found out: how organisms operate, and why they work in that way. But knowing what other people have learnt in the past is not enough: you also have to be able to find things out for yourself, and so you have to learn how to become a researcher. Nowadays, almost all research is quantitative, so no biologist's education is complete without a training in how to take measurements, and how to use the measurements you have taken to answer biological questions.

By the time you have reached advanced level, you will no doubt already have had to undertake a research project, collected results and analysed them in some way. However, you were probably not really sure *why* you had to do what you did. This opening chapter brings up the sorts of questions that you might have worried about, and attempts to answer them. Hopefully it will help you understand why you should bother learning about the world of quantitative biology and statistics. The chapter ends by introducing the subject of how to choose the correct statistical tests for your research project.

1.2 Awkward questions

The first thing you are invariably told to do when carrying out a research project is to make repeated measurements: to include tens or even hundreds of people in surveys; or to have large numbers of replicates in experiments. This seems to be a great deal of wasted effort, so the first question that this book needs to answer is **why do biologists have to repeat everything?**

You are then told to subject your results to statistical analysis. Unfortunately, few subjects are less inviting to most biology students than statistics. For a start it is a branch of mathematics - not usually a biologist's strong suit. You might feel that as you are studying biology you should be able to leave the horrors of maths behind you. So the second question that any book on biological statistics needs to answer is **why do biologists have to bother with statistics?**

Many students also have a problem with the ideas behind statistics. You might well have found that statisticians seem to think in a weird inverted kind of way that is at odds with normal scientific logic. So this book also has to answer the question **why is statistical logic so strange?**

Finally, students often complain, not unreasonably, about the size of statistics books and the amount of information they contain. The reason for this is that there are large numbers of statistical tests, so this book also needs to answer the question **why are there so many different statistical tests?**

In this opening chapter I hope that I can answer these questions and so help put the subject into perspective and encourage you to stick with it. This chapter can be read as an introduction to the information which is set out in what I hope is a logical order throughout the book; it should help you work through the book, either in conjunction with a taught course, or on your own. For those more experienced and confident about statistics, and in particular those with an experiment to perform or results to analyse, you can go directly to the **decision chart for simple statistical tests** (Figure 1.1) introduced later in this chapter on page 7 and also inside the back cover of the book. This will help you choose the statistical test you require and direct you to the instructions on how to perform each test, which are given later in the book. Hence the book can also be used as a handbook to keep around the laboratory and consult when required.

1.3 Why do biologists have to repeat everything?

Why do biologists have to repeat everything when they are conducting surveys or analysing experiments? After all, physicists don't need to do it when they are comparing the masses of sub-atomic particles. Chemists don't need to when they're comparing the pHs of different acids. And engineers don't need to when they are comparing the strength of different shaped girders. They can just generalise from single observations; if a single neutron is heavier than a single proton, then that will be the case for all of them.

However, if you decided to compare the heights of fair- and dark-haired women it is obvious that measuring just one fair-haired and one dark-haired woman would be stupid. If the fair-haired woman was taller, you couldn't generalise from these single observations to tell whether fair-haired women are *on average* taller than dark-haired ones. The same would be true if you compared a single man and a single woman, or one rat that had been given growth hormone with another that had not. Why is this? The answer is, of course, that in contrast to sub-atomic particles, which are all the same, people (in common with other organisms, organs and cells) are all different from each other. In other words they show **variability**, so no one person or cell or experimentally treated organism is typical. It is to get over the problem of variability that biologists have to do so much work and have to use statistics.

To overcome variability, the first thing you have to do is to make **replicated observations** of a **sample** of all the observations you could possibly make. There are two ways in which you can do this.

1. You can carry out a **survey**, sampling at random from the existing **population** of people or creatures or cells. You might measure 20 fair-haired and 20 dark-haired women, for instance.
2. You can create your own samples by performing an **experiment**. Your experimental subjects are then essentially samples of the infinite population of subjects that you *could* have created if you had infinite time and resources. You might, for instance, perform an experiment in which 20 **experimentally treated** rats were injected with growth hormone and 20 other **controls** were kept in exactly the same way except that they received no growth hormone.

1.4 Why do biologists have to bother with statistics?

At first glance it is hard to know exactly what you should do with all the observations that you make, given that all creatures are different. This is where statistics comes in; it helps you deal with the variability. The first thing it helps you do is to examine exactly how your observations vary, in other words to investigate the **distribution** of your samples. The second thing it helps you do is calculate reasonable **estimates** of the situation in the whole population, for instance working out how tall the women are *on average*. These estimates are known as **descriptive statistics.** How you do both of these things is described in Chapter 2.

Descriptive statistics summarise what you know about your samples. However, few people are satisfied with simply finding out these sorts of facts; they usually want to answer questions. You would want to know whether one group of the women was on average taller than the other, or you might want to know whether the rats that had been given the growth hormone were heavier than those which hadn't. **Hypothesis testing** enables you to answer these questions. If you compared the groups, you would undoubtedly find that they were at least slightly different (let's say the fair-haired women were taller than the dark-haired) but there could be two reasons for this. It could be because there really is a difference in height between fair- and dark-haired women. However, it is also possible that you obtained this difference *by chance* by virtue of the particular people you chose. To discount this possibility, you would have to carry out a **statistical test** (in this case a two-sample t test) to work out the probability that the apparent effects *could* have occurred by chance. If this probability was small enough you could make the judgement that you could discount it and decide that the effect was **significant**. In this case you would then have decided that fair-haired women are *significantly taller* than dark-haired.

1.5 Why is statistical logic so strange?

All of this has the consequence that the logic of hypothesis testing is rather counterintuitive. When you are investigating a subject in science, you typically make a hypothesis that something interesting is happening, for instance in our case that fair-haired women are taller than dark-haired, and then set out

to test it. In statistical hypothesis testing you do the opposite. You construct a **null hypothesis** that *nothing interesting* is happening, in this case that fair- and dark-haired women have the same mean height, and then test whether this null hypothesis is true. Statistical tests have four main stages.

null hypothesis A preliminary assumption in a statistical test that the data shows no differences or associations. A statistical test then works out the probability of obtaining data similar to your own by chance.

Step 1: Formulating a null hypothesis

The null hypothesis you must set up is the opposite of your scientific hypothesis: that there are no differences or relationships. (In the case of the fair- and dark-haired women, the null hypothesis is that they are the same height.)

Step 2: Calculating a test statistic

The **test statistic** you calculate measures the size of any effect (usually a difference between groups or a relationship between measurements) relative to the amount of variability there is in your samples. Usually (but not always) the larger the effect, the larger the test statistic.

Step 3: Calculating the significance probability

Knowing the test statistic and the size of your samples, you can calculate the probability of getting the effect you have measured, just by chance, *if the null hypothesis were true.* This is known as the **significance probability**. Generally the larger the test statistic and sample size, the smaller the significance probability.

significance probability The chances that a certain set of results could be obtained if the null hypothesis were true.

Step 4: Deciding whether to reject the null hypothesis

The final stage is to decide whether to reject the null hypothesis or not. By convention it has been decided that you can reject a null hypothesis if the significance probability is less than or equal to 1 in 20 (a probability of 5% or 0.05). If the significance probability is greater than 5%, you have no evidence to reject the null hypothesis – *but this does not mean you have evidence to support it.*

The 5% cut-off is actually something of a compromise to reduce the chances of biologists making mistakes about what is really going on. For instance, there is a 1 in 20 chance of finding an apparent significant effect, even if there wasn't a real effect. If the cut-off point had been lowered to, say, 1 in 100 or 1%, the chances of making this sort of mistake (known to statisticians as a **type 1 error**) would be reduced. On the other hand, the chances of failing to detect a real effect (known as a **type 2 error**) would be increased by lowering the cut-off point.

type 1 error The detection of an apparently significant difference or association, when in reality there is no difference or association between the populations.

type 2 error The failure to detect a significant difference or assocation, when in reality there is a difference or association between the populations.

As a consequence of this probabilistic nature, performing a statistical test does not actually allow you to *prove* anything conclusively. If your test tells you there is a significant effect, there is still a small chance that there might not really have been one. Similarly, if your test is not significant, there is still a chance that there might really have been an effect.

1.6 Why are there are so many statistical tests?

Even if we accept that statistical tests are necessary in biology, and can cope with the unusual logic, it is perhaps not unreasonable to expect that we should be able to analyse all our results using just a single statistical test. However,

3.3 What to do if your data has a significantly different distribution from the normal

3.3.1 Symmetrically distributed data

If the data are not normally distributed but are symmetrically distributed around the mean (Figure 2.2a), you don't need to do anything. It is generally thought valid to use parametric tests. This is true of the data on bull elephants we have just examined. This is because, according to the central limit theorem, the means of samples which are symmetrically distributed tend to be normally distributed if the sample size is large enough.

3.3.2 Skewed data

transformation
A mathematical function used to make the distribution of data more symmetrical and so make parametric tests valid.

If the data are asymmetrically distributed, either being **positively skewed** (Figure 2.2b) or **negatively skewed** (Figure 2.2c), you may need to use a **transformation** to make the distribution more symmetrical before you do any statistical analysis.

Positively skewed data may be transformed into a more symmetrical distribution in several ways, all of which separate the smaller numbers. Possible transformations of a variable x are to take $\sqrt{x}$, $\log_{10} x$, or $1/x$.

To transform negatively skewed data into normally distributed data, you need to separate the larger numbers. Possible transformations are powers of x like x^2, x^3 or even higher powers of x.

3.3.3 Irregularly distributed samples and ranked data

median
The central value of a distribution (or average of the middle points if the sample size is even).

non-parametric test
A statistical test which does not assume that data is normally distributed, but instead uses the ranks of the observations.

parametric test
A statistical test which assumes that data are normally distributed.

If the data is irregularly distributed, as is commonly the case with many ecological and behavioural surveys (Figure 2.2d), there is no sensible transformation you can use to make the data normally distributed. This is also true of two other sorts of data: **discretely** varying data, particularly when the number of possible states is low (say, 1, 2, 3 or 4); and **ranked** data (like pecking order or degree of patterning) which may be all that ecologists, behavioural scientists or psychologists can collect, or all the output that you can get from a questionnaire.

In these cases, it is inappropriate to describe your data by giving its mean and standard deviation. Instead it is better to present its **median** value and **interquartile range**, showing the results graphically with a box and whisker plot. In these cases when carrying out statistical analysis, you will also have to use the **non-parametric** tests given in the next three chapters rather than the **parametric tests.**

3.3.4 Proportional data

In some cases there are theoretical reasons why your data could not possibly be normally distributed. For instance, **proportional data,** such as the ratio of root to total plant mass, cannot be normally distributed because values cannot fall

Using MINITAB

Click on the **Stat** menu; from there move onto the **Basic Statistics** bar, then click on **Normality Test.** This brings up the **Normality Test** window (below). Click the mouse in the **Variable** box, highlight **C1 bulls** and click on **Select.** This will put **C1 bulls** in the **Variable** box. Next tick **Kolgomorov–Smirnov** to give the completed window shown below.

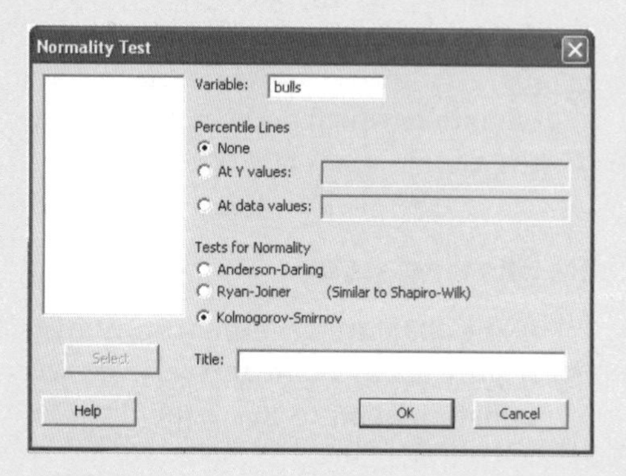

Finally click on OK to run the test. MINITAB will give you the following plot which presents the Kolgomorov–Smirnov statistic (K-S). Here it is 0.172.

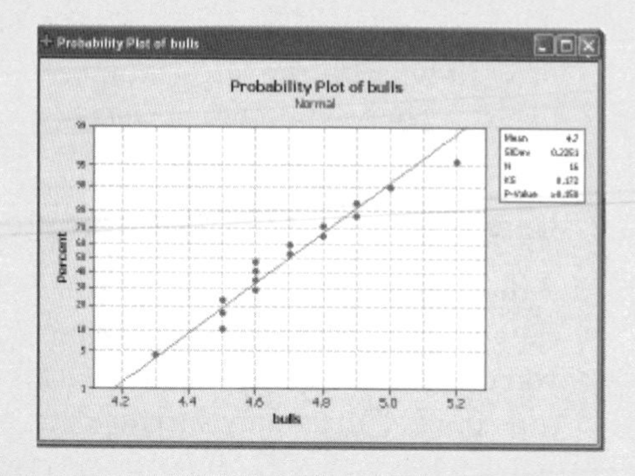

Step 3: Calculating the significance probability

SPSS calls the significance probability Sig. and gives it as 0.200. MINITAB calls the significance probability *P*-value and gives it as >0.15.

Step 4: Deciding whether to reject the null hypothesis

If Sig. or *P*-value ≤ 0.05, you must reject the null hypothesis. Therefore you can say that the distribution is significantly different from the normal distribution.

If Sig.or *P*-value > 0.05, you have no evidence to reject the null hypothesis. Therefore you can say that the distribution is not significantly different from the normal distribution.

Here Sig. and *P*-value > 0.05, so you have no evidence to reject the null hypothesis. The distribution of the weights of the bull elephants is not significantly different from the normal distribution. **You can analyse this data using parametric tests!**

Example 3.1

Is the distribution of weights of the bull elephants significantly different from the normal distribution?

Solution

Step 1: Formulating the null hypothesis

The null hypothesis is that the distribution of your sample is *not* different from the normal distribution.

Here the null hypothesis is that the distribution of the weights of the bull elephants is *not* different from the normal distribution.

Step 2: Calculating the test statistic

Using SPSS

Go into the **Analyse** menu and choose **Descriptive Statistics** and then click on **Explore.** Set the program to explore the bull weight data, but before running the test, click on **Plots** and in the **Explore Plots** dialogue box tick **Normality plots with tests**. The completed box is shown below.

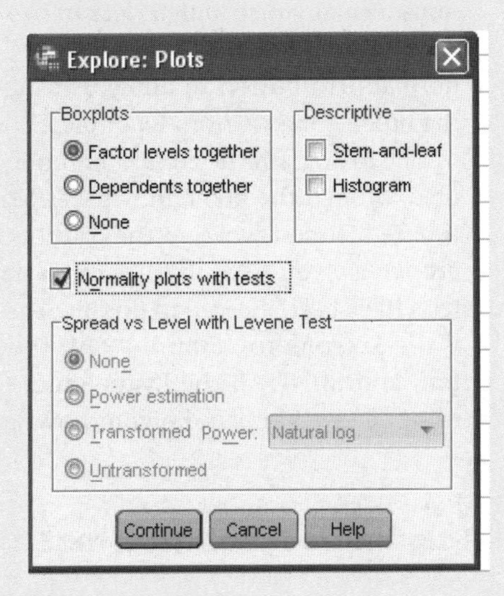

Finally click on **Continue** and then **OK** to run the test. SPSS will give you not only the usual descriptive statistics but also an extra table and two more plots. The important results are shown in the table below.

Tests of normality

	Kolmogorov-Smirnov[a]			Shapiro-Wilk		
	Statistic	df	Sig.	Statistic	df	Sig.
Bulls	0.172	16	0.200*	0.962	16	0.691

[a]Lilliefors significance correction

*This is a lower bound of the true significance.

The table gives the Kolgomorov–Smirnov statistic. Here it is 0.172.

3 Testing for normality and transforming data

3.1 The importance of normality testing

We saw in Chapter 1 that it is very important to determine whether your data is normally distributed. If it is normally distributed, you can use a series of very powerful statistical tests, parametric tests, that make this assumption. However, non-normally distributed data has to be analysed using the less powerful non-parametric tests. But how can we tell whether our data is close enough to the normal distribution to allow us to use parametric tests? Let's use our data for the bull elephants from Example 2.1 as an example.

The histogram the computer produces does not look exactly normally distributed, though the box and whisker plots show that it is fairly symmetrically distributed. To test whether the distribution is *significantly different from the normal distribution* you need to carry out a statistical test. Fortunately, several statistical tests have been devised to do this. The one that both SPSS and MINITAB perform is the Kolgomorov–Smirnov test, though SPSS also performs the Shapiro–Wilk test, and MINITAB the Ryan–Joiner test and Anderson–Darling test. I suggest using the well-known Kolgomorov–Smirnov test.

3.2 The Kolgomorov–Smirnov test

3.2.1 Purpose

To test whether the distribution of a sample is significantly different from the normal distribution.

3.2.2 Carrying out the test

The method of carrying out a Kolgomorov–Smirnov test is best illustrated by using as an example the weights of the bull elephants whose distribution we have just examined. Like all statistical tests it has four steps.

below 0 or rise above 1, so the distribution is cut off at both ends. Before dealing with proportional data, therefore, each data point x has to be transformed by calculating arcs in $\sqrt{x}$.

3.4 Examining data in practice

To illustrate how you should investigate the distribution of your data and deal with it if it does not look normally distributed, it is perhaps best to go through two examples.

Example 3.2

Data on the masses of 30 newts gave the following results.

Mass (g)

0.91	1.45	5.32	2.97	2.12	0.76	1.85	1.53	2.42	1.92
1.86	3.81	6.54	2.53	1.92	2.33	1.45	1.22	1.68	3.10
1.80	3.51	2.43	1.34	1.09	2.62	1.90	4.32	0.89	1.55.

Test to see if this data is normally distributed and, if not, carry out a sensible transformation to make it normally distributed.

Solution

Using SPSS

SPSS gives the following histogram:

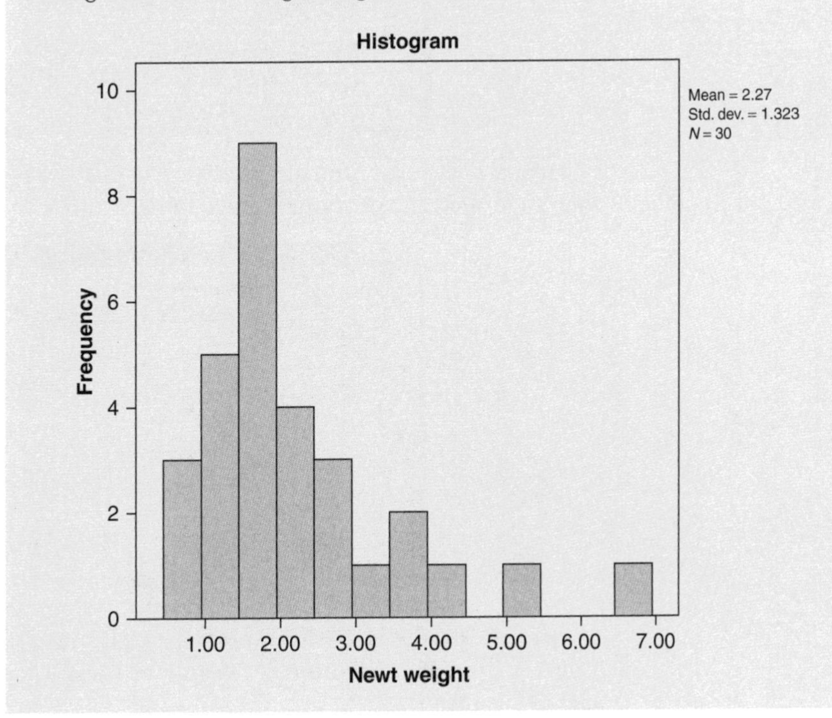

It is clearly positively skewed. Performing a normality test gives the following results:

Tests of normality

	Kolmogorov–Smirnov[a]			Shapiro–Wilk		
	Statistic	df	Sig.	Statistic	df	Sig.
Newt weight	0.177	30	0.018	0.857	30	0.001

[a]Lilliefors' significance correction

Here, Sig. is 0.018, so the data is distributed significantly differently from the normal distribution. It needs to be transformed strongly, ideally by taking logarithms.

Using MINITAB

Examining the distribution in MINITAB gives the following histogram:

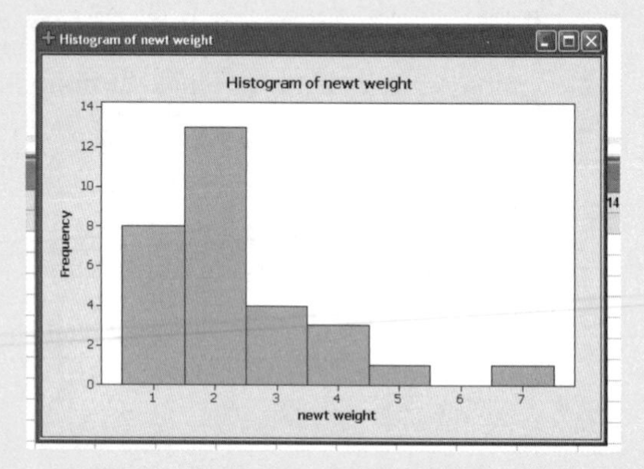

The data is clearly strongly positively skewed with far more small newts than large ones, and performing a test for normality gives the following results.

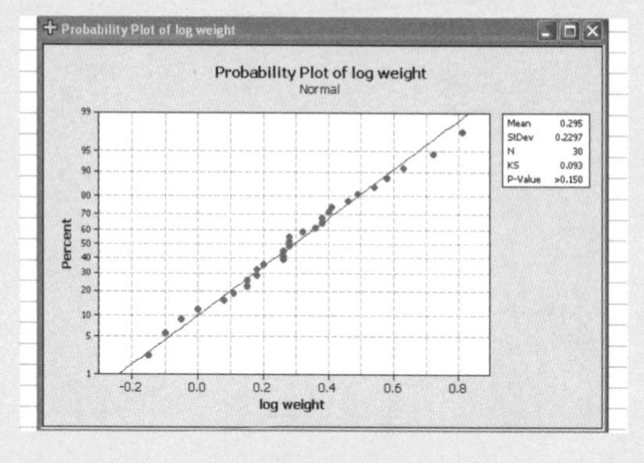

Here, the *P*-value is 0.015, so the data is distributed significantly differently from the normal distribution. It needs to be transformed strongly, ideally by taking logarithms.

3.5 Transforming data

3.5.1 Using SPSS

To transform data, click on the **Transform** bar, then click on the **Compute** option. This brings up the **Compute Variable** dialogue box. A new column (lognewt) can then be created. Type lognewt in the **target variable** box and put the expression **Lg10(newtweight)** into the **numeric expression** box. Do this by first clicking on **Arithmetic** from the **Function group** box and then Lg10 from the **Functions and Special Variables** box. Move the choice into the **Numeric Expression** box by clicking on the upward-pointing arrow. Finally click onto **newtweight** and the right-pointing arrow to end up with the **Numeric Expression** LG10(newtweight). The completed dialogue box is shown below.

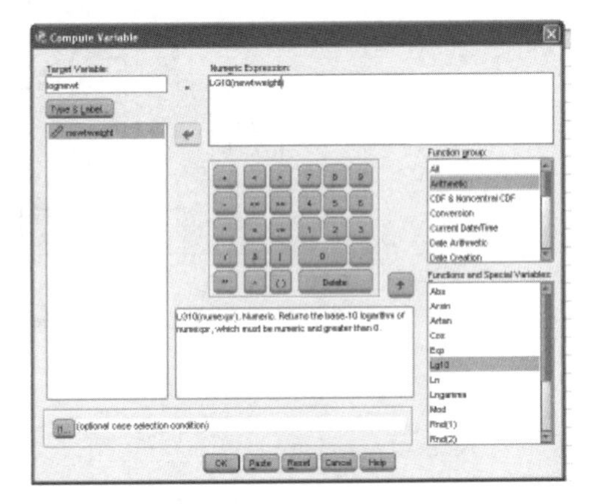

Click on **OK** and SPSS will produce the new column **lognewt** next to **newtweight** as seen below.

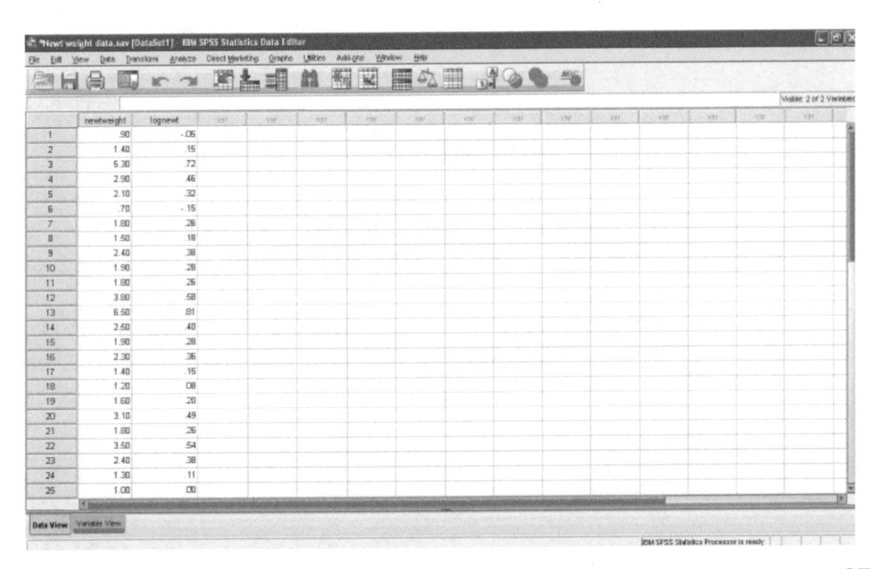

	newtweight	lognewt
1	.90	-.05
2	1.40	.15
3	5.30	.72
4	2.90	.46
5	2.10	.32
6	.70	-.15
7	1.80	.26
8	1.50	.18
9	2.40	.38
10	1.90	.28
11	1.80	.26
12	3.80	.58
13	6.50	.81
14	2.50	.40
15	1.90	.28
16	2.30	.36
17	1.40	.15
18	1.20	.08
19	1.60	.20
20	3.10	.49
21	1.80	.26
22	3.50	.54
23	2.40	.38
24	1.30	.11
25	1.00	.00

You can now examine the distribution of $\log_{10}$(newt weight) using SPSS and carry out a further Kolgomorov–Smirnov test. It will give the following results.

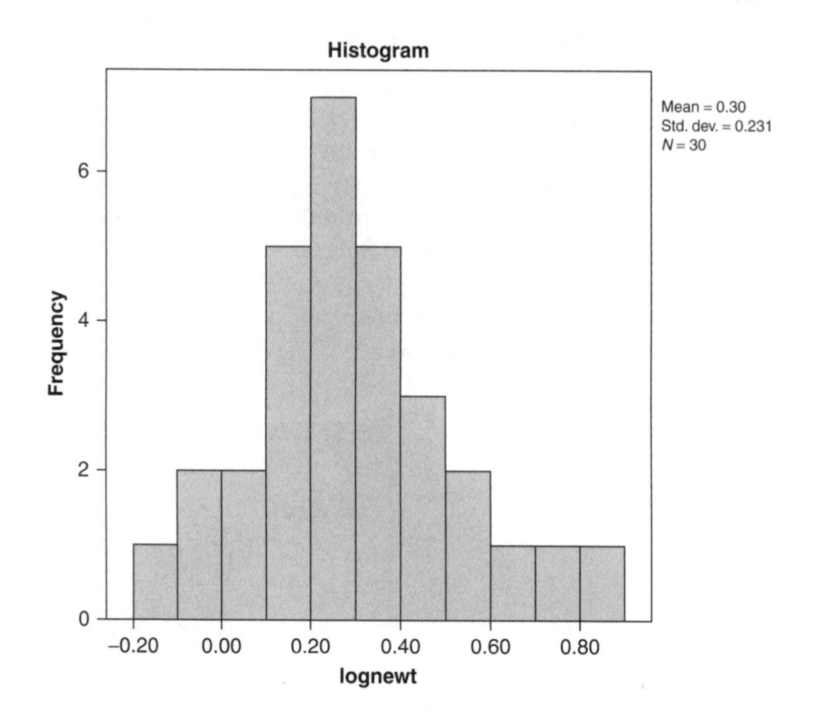

Tests of normality

	Kolmogorov–Smirnov[a]			Shapiro–Wilk		
	Statistic	df	Sig.	Statistic	df	Sig.
lognewt	0.095	30	0.200*	0.987	30	0.972

[a]Lilliefors' significance correction.

*This is a lower bound of the true significance.

3.5.2 Using MINITAB

To transform data, click on the **Calc** bar, then click on the **Calculator** option. This brings up the **Calculator** dialogue box. A new column (log weight) can then be created. Do this by first clicking on Log Base 10 from the **Function group** box and click on **Select**. This puts LOGTEN(number) into the **Expression** box. Then click onto C1 newt weight and click onto **Select**. This makes the **Expression** box read LOGTEN ('newt weight'). Then type C2 into the **Store result in variable:** box. The completed dialogue box is shown below.

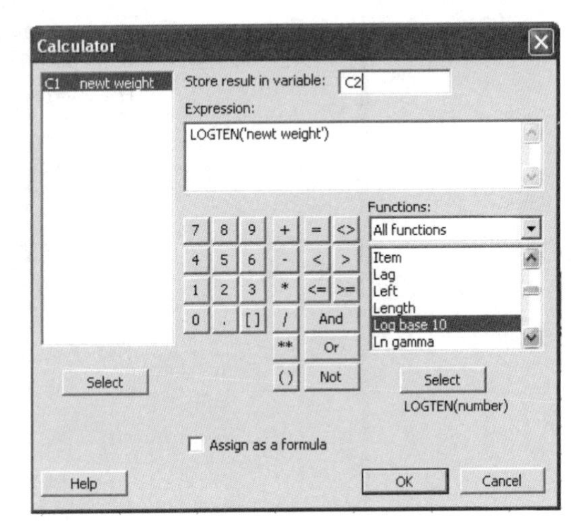

Click on **OK** and MINITAB will produce the new column C2, which you can name **log weight** as seen below.

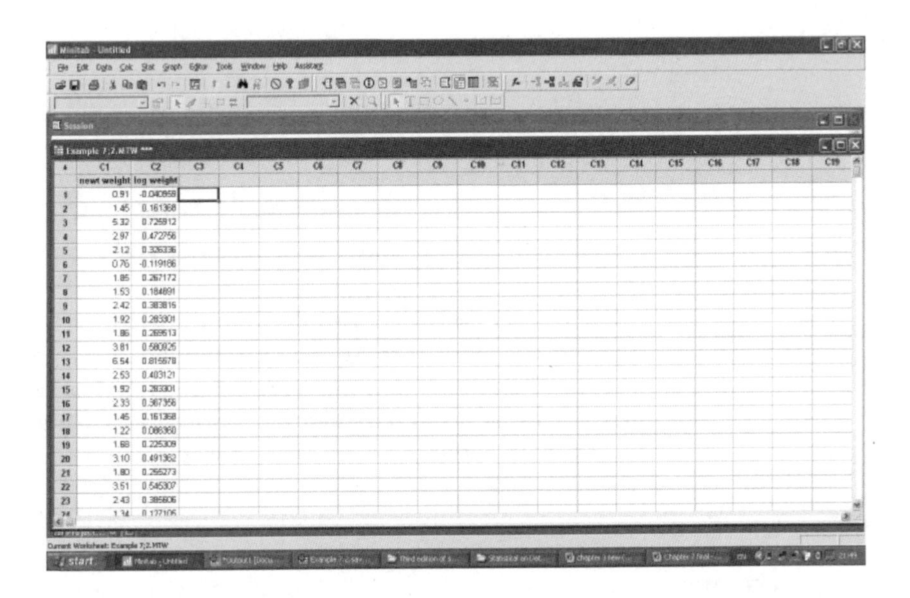

You can now examine the distribution of log weight and carry out a Kolgomorov–Smirnov test. It will give the following results.

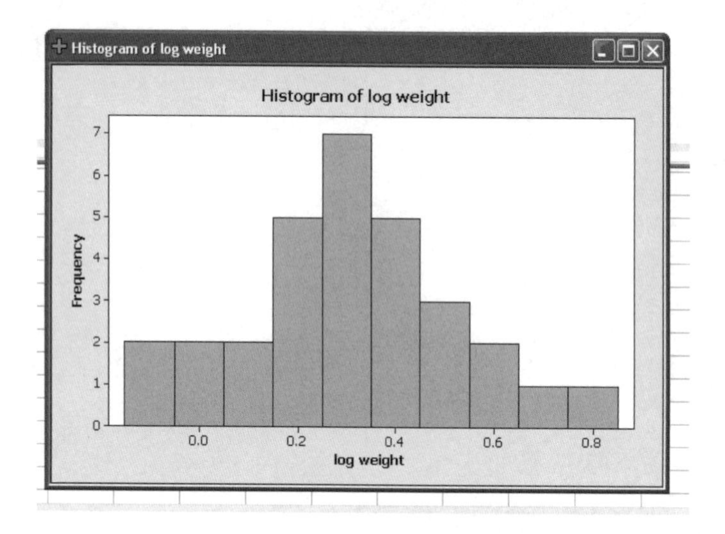

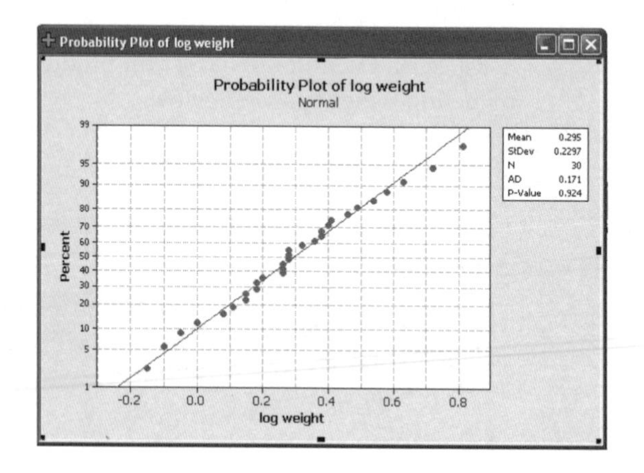

Whichever package is used the new histogram is far more symmetrical. Here Sig. and P-value > 0.05, so the distribution is *not* significantly different from normal. This shows that the log transformation has been successful. **It is now possible to carry out parametric statistical tests using this transformed data.**

Let's have a look at another example as a contrast.

Example 3.3

In a survey to investigate the population of ground beetles in a field of wheat, 20 pitfall traps were set out through the field, and samples collected after 3 days. The numbers of ground beetles found in each trap were as follows:

12	15	0	2	26	0	1	18	3	0
0	5	17	0	13	1	10	2	8	13

Examine the distribution of the data. Is it normally distributed or could it be transformed to be so? Calculate relevant descriptive statistics for the results.

Solution

Exploring this data in MINITAB and testing for normality gives the following relevant statistical results, histogram and box and whisker plots:

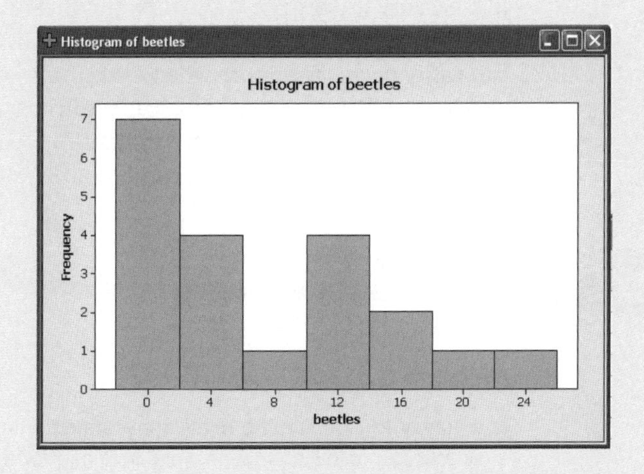

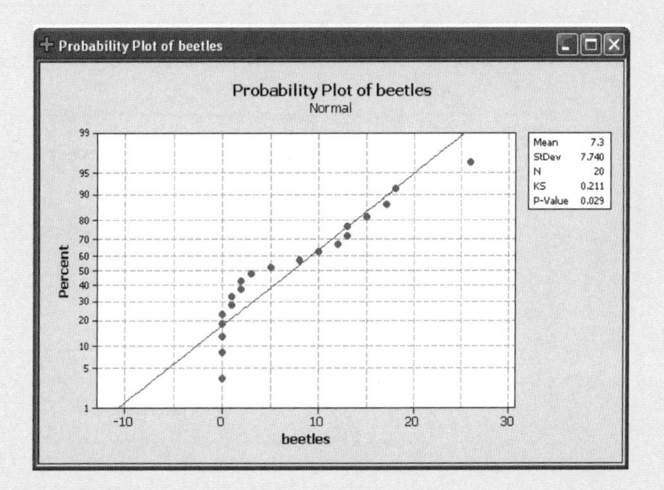

It is clear from looking at the histogram that the distribution is very irregular, with two separate peaks in the distribution. The Kolgomorov–Smirnov test gives a *P*-value of 0.029, so it is significantly different from the normal distribution. It is therefore inappropriate to describe the data using its mean and standard deviation. Because the data is so irregular and there are so many zeros, it would also be impossible to transform it to make it normally distributed. Instead it is best described by its median (4) and interquartile range (12.75), and by showing a box and whisker plot.

3.6 The complete testing procedure

The complete process is best summarised by the flow chart in Figure 1.2, presented on page 8 and on the inside cover of the book. One warning is in order, however. The Kolgomorov–Smirnov test is useful, but you must remember that just because it does not show up a significant difference *does not mean that your data is normally distributed.* Particularly with small sample sizes, a type 2 error can easily occur. Therefore when examining the distribution of data you should also look at histograms as well. The results of the Kolgomorov–Smirnov tests are useful, however, when you are presenting your results to back up claims that a parametric analysis you have carried out is valid.

3.7 Self-assessment problems

Problem 3.1

In a survey to examine the relative investment into roots of a self-supporting plant and a climber, the following results were obtained for the proportion of total dry mass in the root system.

Self-supporting	0.16	0.23	0.28	0.22	0.25	0.20	0.17	0.32	0.24	0.26
Climber	0.15	0.13	0.08	0.11	0.13	0.19	0.14	0.16	0.15	0.24
	0.13	0.07.								

How should this data be transformed to make it normally distributed?

Problem 3.2

A survey was carried out of the mean lengths of 20 species of the crow family. The results are shown below.

Species	1	2	3	4	5	6	7	8	9	10
	11	12	13	14	15	16	17	18	19	20
Length	7	8	9	10	11	12	12	13	14	15
	16	18	19	21	23	24	26	29	32	37

Examine the data using SPSS or MINITAB. Does it look normally distributed and if not how would you transform it to make it so?

4 Testing for differences between one or two groups

4.1 Introduction

We saw in Chapter 2 how we can summarise the distribution of a sample and in Chapter 3 how we can tell whether that distribution is normal or not. These are useful things, but hardly satisfying. If we are going to carry out biological research, we often want to answer specific *questions* about the things we are measuring. In particular there are several *comparisons* that we might want to make:

- We might want to *compare* the average value of a measurement taken on a single population with an expected value. Is the birth weight of babies from a small town *different* from the national average?
- We might want to *compare* two sets of related measurements or paired measurements made on a single population. Are male gibbons heavier than their mates? Do patients have a *different* heart rate before and after taking beta blockers? Or is the pH of ponds *different* at dawn and dusk?
- We might want to *compare* experimentally treated organisms or cells with controls. Do shaken sunflowers have *different* mean height from unshaken controls?
- We might want to *compare* two groups of organisms or cells. Do different strains of bacteria have *different* growth rates?

***t* tests**
Statistical tests which analyse whether there are differences between measurements on a single population and an expected value, between paired measurements, or between two unpaired sets of measurements.

The first half of this chapter describes how you can use a simple set of tests called ***t* tests** to determine whether there are differences and how to work out how big those differences are. The second half shows you how to do the same thing using non-parametric equivalents if you have ranked data or if your measurements are not normally distributed.

4.2 Why we need statistical tests for differences

4.2.1 The problem

You might imagine it would be easy to find out whether there are differences. You would just need to take measurements on your samples and compare the average values to see if they were different. However, there is a problem. Because of variation we can never be certain that the differences between our **sample means** reflect real differences in the **population means**. We might have got different means just by chance.

4.2.2 The solution

Suppose μ is the mean length of a population of rats. If we take measurements on a sample of rats, it is quite likely we could get a mean value $\bar{x}$ that is one standard error $\overline{SE}$ greater or smaller than μ. In fact the chances of getting a mean that different *or more* from μ is equal to the shaded area in Figure 4.1a. In contrast it is much less likely that we could get a value that is different by more than three standard errors from μ (Figure 4.1b). The probability is given by the tiny (though still real) area in the two tails of the distribution.

There is a point, usually around or just above two standard errors, where the probability of getting a mean that is different or more by chance, will fall below 5%. Therefore if a sample mean is more than this different from the expected mean, we can say that it is **significantly different**. However, because of variability, we cannot be *sure* that these differences are real.

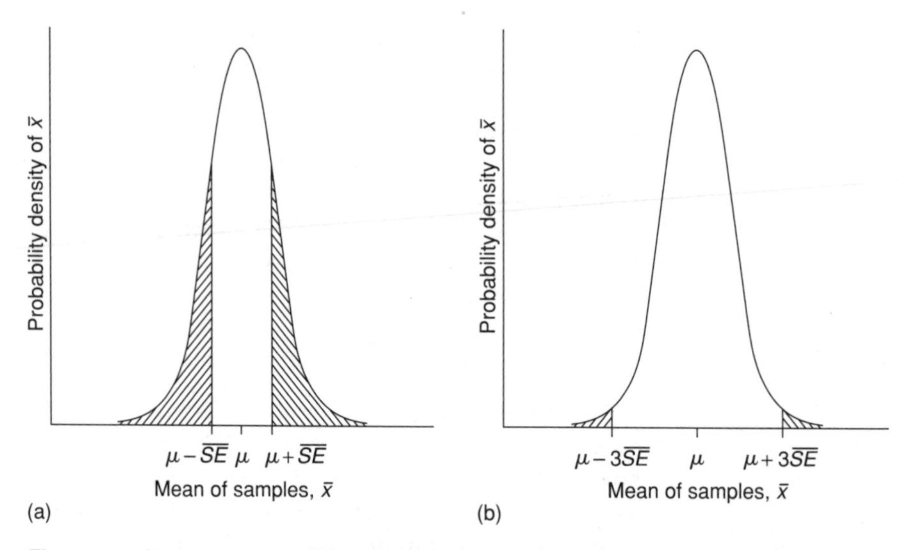

Figure 4.1 Sample means different from an expected value. (a) There is a high probability (shaded areas) of obtaining a mean at least one standard error $\overline{SE}$ away from the expected mean μ (b) There is a very low probability (shaded areas) of getting a mean at least three standard errors 3 $\overline{SE}$ away from the expected mean μ.

4.3 How we test for differences

As we saw in Chapter 1, carrying out statistical tests involves a somewhat inverted form of logic that has four main stages. The stages involved in carrying out *t* tests for differences are shown below.

null hypothesis
A preliminary assumption in a statistical test that the data shows no differences or associations. A statistical test then works out the probability of obtaining data similar to your own by chance.

Step 1: Formulating a null hypothesis

The first stage is to assume the opposite of what you are testing. Here, as you are testing whether there is a difference, the **null hypothesis** is that there is **no** difference.

Step 2: Calculating the test statistic

The next stage is to examine the data values and calculate a test statistic from them. When testing for differences, the test statistic is usually a measure of how different the means are relative to the variability. The greater the difference in the means and the smaller the scatter in the data, the bigger the absolute value of the test statistic, t (i.e. the further away from zero it will be). The smaller the difference in the means and the greater the scatter, the smaller the absolute value of the test statistic.

Step 3: Calculating the significance probability

Next you must examine the test statistic, t, and assess the probability of getting an absolute value that high or greater if the null hypothesis were true. The larger the absolute value of the test statistic (i.e. the further away from zero it is), hence the greater the distance between the means, the smaller the probability will be. The smaller the absolute value of the test statistic, the larger the probability will be.

significant difference
A difference which has less than a 5% probability of having happened by chance.

Step 4: Deciding whether to reject the null hypothesis

- If the significance probability is below a cut-off point, you must reject the null hypothesis and conclude that there is a **significant difference**. As we have seen, usually in biology one rejects the null hypothesis if the significance probability is less than or equal to 1 in 20. This probability is often written as the decimal 0.05, or as 5%. This criterion for rejecting the null hypothesis is therefore known as the 5% significance level.
- If the significance probability is greater than 5%, you have no evidence to reject the null hypothesis. But this does not mean you have evidence to support it.

critical values
Tabulated values of test statistics; usually if the absolute value of a calculated test statistic is greater than or equal to the appropriate critical value, the null hypothesis must be rejected.

Statisticians have taken a lot of the hard work out of deciding whether to reject the null hypothesis by preparing tables of **critical values** for test statistics. Several of these tables, including the one for the t statistic, are given in the tables at the end of the book. If the absolute value of your test statistic is (usually) greater than or equal to the critical value for the 5% significance level, then there is a less than 5% probability of getting these results by chance. Therefore, you can reject the null hypothesis. It is even easier if you are carrying out a statistical test in SPSS or another computer package. It will work out the significance probability for you, and all you have to do is compare that probability with 0.05.

Sometimes you may find the probability P falls below critical levels of 1 in 100 or 1 in 1000. If this is true, you can reject the null hypothesis at the 1% or 0.1% levels respectively.

confidence limits
Limits between which estimated parameters have a defined likelihood of occurring. It is common to calculate 95% confidence limits, but 99% and 99.9% confidence limits are also used. The range of values between the upper and lower limits is called the confidence interval.

Step 5: Calculating confidence limits

Whether or not there is a **significant difference**, you can calculate **confidence limits** to give a set of plausible values for the differences of the means. Calculating

95% confidence limits for the difference of means is just as straightforward as calculating 95% confidence limits for the means themselves (Section 2.6).

4.4 One- and two-tailed tests

two-tailed tests
Tests which ask merely whether observed values are different from an expected value or each other, not whether they are larger or smaller.

Statistical tables often come in two different versions: one-tailed and two-tailed. Most biologists use **two-tailed tests**. These test whether there are differences from expected values but do not make any presuppositions about which way the differences might be. With our rats, therefore, we would be testing whether they had a different length but not whether they were longer or shorter than expected. The criterion for rejecting the null hypothesis in the two-tailed test is when the total area in the two tails of the distribution (Figure 4.1) is less than 5%, so each tail must have an area of less than 2.5%. All the statistical tables in this text are two-tailed. The tests carried out by SPSS and MINITAB are also by default two-tailed.

4.5 The types of *t* test and their non-parametric equivalents

There are three main types of *t* test, which are used in different situations. The simplest one, the **one-sample *t* test**, is used to determine whether the mean of a single sample is different from an expected value. If you want to see if there are differences between two sets of paired observations, you need to use the **paired *t* test**. Finally to test whether the means of two independent sets of measurements are different you need to carry out a **two-sample *t* test**, also known as an **independent sample *t* test**. These tests all have fairly easy to grasp logic and are straightforward to carry out mathematically. Therefore, instructions will be given to carry out these tests both using a calculator, and the computer packages.

parametric test
A statistical test which assumes that data are normally distributed.

All these *t* tests are so-called **parametric tests**, which assume that the data is normally distributed. If you have found that your data is not normally distributed (see Section 3.3), and cannot transform it into data that is, you should instead use their **non-parametric** equivalents: the **sign test**, the **Wilcoxon matched pairs test** and the **Mann–Whitney *U* test**. These are given at the end of the chapter.

4.6 The one-sample *t* test

4.6.1 Purpose

To test whether the sample mean of one measurement taken on a single population is different from an expected value E.

4.6.2 Rationale

The one-sample *t* test calculates how many standard errors the sample mean is away from the expected value. It is therefore found using the formula

$$t = \frac{\text{Sample mean} - \text{Expected value}}{\text{Standard error of mean}} = \frac{\bar{x} - E}{\overline{SE}} \qquad (4.1)$$

The further away the mean is from the expected value, the larger the value of *t*, and the less probable it is that the real population mean *could* be the expected value. Note, however, that it does not matter whether *t* is positive or negative. It is the *difference* from zero that matters, so you must consider the absolute value of *t*, $|t|$. If $|t|$ is greater than or equal to a critical value, then the difference is significant.

4.6.3 Validity

The data must be normally distributed.

4.6.4 Carrying out the test

To see how to carry out a test in practice it is perhaps best to run through a straightforward example.

Example 4.1

Do the bull elephants we first met in Example 2.1 have a different mean mass from the mean value for the entire population of African elephants of 4.50 tonnes?

Solution

Step 1: Formulating the null hypothesis

The null hypothesis is that the mean of the population *is not* different from the expected value. Here, therefore, the null hypothesis is that the mean weight of bull elephants is 4.50 tonnes.

Step 2: Calculating the test statistic

Using a calculator

Here, the mean weight $\bar{x}$ of the sample of bull elephants is 4.70 tonnes with an estimate of the standard error $\overline{SE}$ of 0.0563 tonnes. Therefore using equation 4.1

$$t = (4.70 - 4.50)/0.0563 = 3.55.$$

The mean is 3.55 standard errors away from the expected value.

Using SPSS

To carry out a one sample *t* test, you first need to enter all the data into a single column, and give the column a name (here **bulls**). To run the test, click on the **Analyze** menu, then move onto the **Compare Means** bar and click on **One-Sample T Test**. SPSS will present you with the **One-Sample T Test** dialogue box.

Put the column you want to compare (here **bulls)** into the **Test Variables** box, and the value you want to compare it (here **4.50**) with in the **Test Value** box. This will give

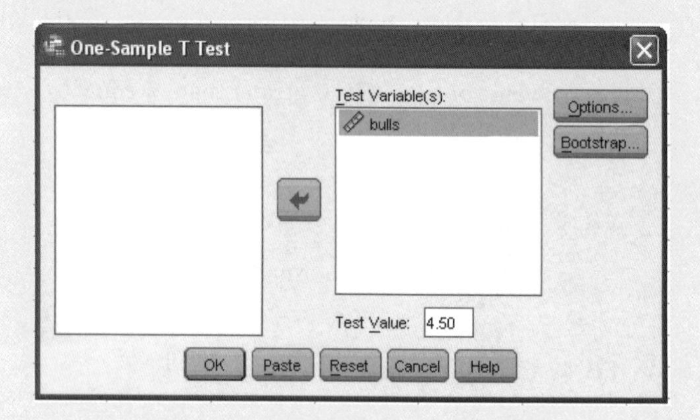

Finally click on **OK** to run the test.

SPSS will produce the following useful tables:

One sample statistics

	N	Mean	Std. deviation	Std. error mean
Bulls	16	4.7000	0.22509	0.05627

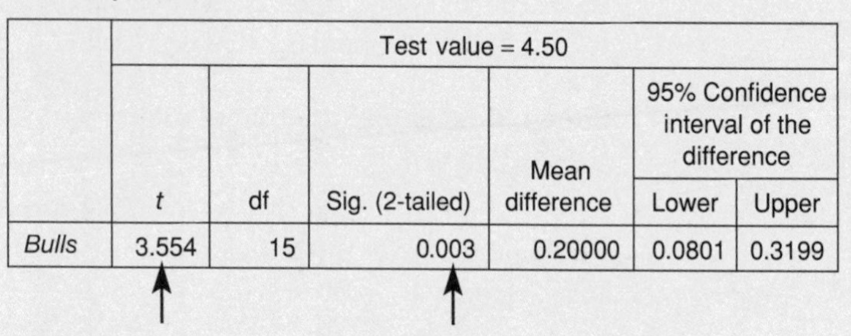

One-sample test

	Test value = 4.50					
					95% Confidence interval of the difference	
	t	df	Sig. (2-tailed)	Mean difference	Lower	Upper
Bulls	3.554	15	0.003	0.20000	0.0801	0.3199

SPSS gives you all the descriptive statistics you need in the top table, and the value of t (3.554) in the lower table.

Using MINITAB

Once again, enter all the data into a single column, and name it **bulls**. To run the test, click on the **Stat** menu, then move onto the **Basic Statistics** bar and click on **1-Sample t.** MINITAB will present you with the **1-Sample t (Test and Confidence Interval)** dialogue box.

Put the column you want to compare (here **bulls**) into the **Samples in columns** box, tick the **Perform hypothesis test,** and type in the value you want to compare (here 4.5 in the **Hypothesised mean** box). This will give

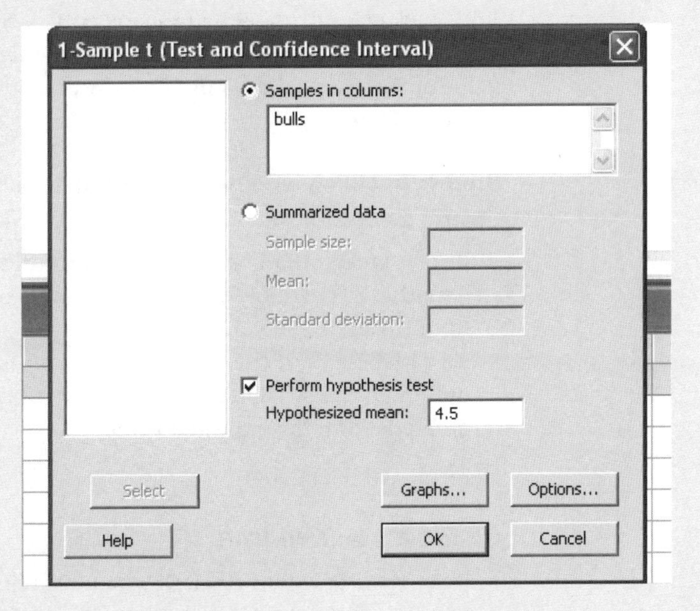

Finally click on **OK** to run the test.

MINITAB will produce the following useful results

One-Sample T: bulls

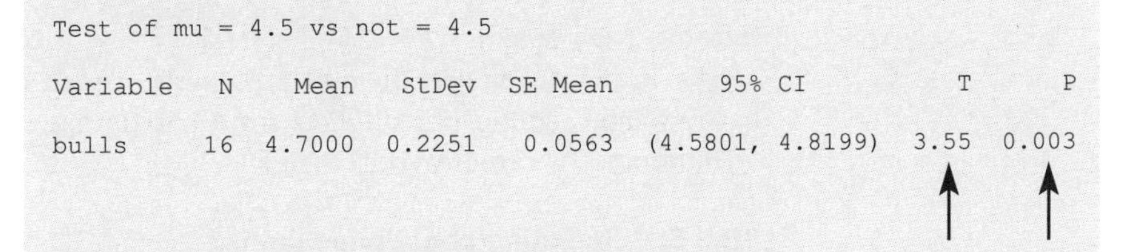

```
Test of mu = 4.5 vs not = 4.5

Variable   N    Mean   StDev  SE Mean        95% CI            T      P

bulls     16  4.7000  0.2251   0.0563  (4.5801, 4.8199)  3.55  0.003
```

giving the descriptive statistics and the value of T, 3.55.

Step 3: Calculating the significance probability

You must calculate the probability P that the absolute value of t, written $|t|$, would be this high or higher if the null hypothesis were true.

Using a calculator

You must compare your value of $|t|$ with the critical value of the t statistic for $(N-1)$ degrees of freedom and at the 5% level $[t_{(N-1)}(5\%)]$. This is given in Table S1 at the end of the book.

Here, there are 16−1 = 15 degrees of freedom, so the critical value that $|t|$ must exceed for the probability to drop below the 5% level is 2.131.

Using SPSS and MINITAB

Both packages calculate the probability, *P*. SPSS calls it **Sig. (two-tailed)** and MINITAB calls it **P** [Note that the bigger the value of $|t|$, the smaller the value of Sig. (two-tailed)] and *P*.

Here Sig. (two-tailed) = $P = 0.003$.

Step 4: Deciding whether to reject the null hypothesis

Using a calculator

- If $|t|$ is greater than or equal to the critical value, you must reject the null hypothesis. Therefore you can say that the mean is significantly different from the expected value.
- If $|t|$ is less than the critical value, you have no evidence to reject the null hypothesis. Therefore you can say that the mean is not significantly different from the expected value.

Here $|t| = 3.55 > 2.131$.

Using SPSS and MINITAB

- If Sig. (two-tailed) or $P \leq 0.05$ you must reject the null hypothesis. Therefore you can say that the mean is significantly different from the expected value.
- If Sig. (two-tailed) or $P > 0.05$ you have no evidence to reject the null hypothesis. Therefore you can say that the mean is not significantly different from the expected value.

Here Sig. (two-tailed) = $P = 0.003 < 0.05$

Therefore we can reject the null hypothesis. We can say that bull elephants have a weight significantly different from 4.50 tonnes; in fact, with a mean of 4.70 tonnes, they are heavier.

Step 5: Calculating confidence limits

Just as you can find confidence intervals for the means of a measurement like weight, so you can also find confidence intervals for the difference between the mean and the expected mean.

Using a calculator

The 95% confidence limits for the difference are given by the equation

$$95\%\ \text{CI(difference)} = \bar{x} - E \pm (t_{(N-1)}(5\%) \times \overline{SE}) \tag{4.2}$$

Here, the mean is 4.70, with a standard error of 0.0563, and the critical *t* value for 15 degrees of freedom is 2.131. Therefore

$$\begin{aligned} 95\%\ \text{CI(difference)} &= \underline{(4.70-4.50)} \pm (2.131 \times 0.0563) \\ &= 0.08\ \text{to}\ 0.32 \end{aligned}$$

Bull elephants are 95% likely to be between 0.08 and 0.32 tonnes heavier than 4.5 tonnes.

Using SPSS and MINITAB

A 95% confidence interval is given by both packages. SPSS gives intervals for the difference in the second table as between 0.0801 to 0.3199. MINITAB gives 95% confidence intervals for the mean itself of 4.5801 to 4.8199. As ever you should not just write these down but round them up to a sensible number of significant figures. Here the difference rounds to 0.08 to 0.32.

4.6.5 Presenting the results

You should present the results of a one-sample *t* test in a way that emphasises the biological reality of what you have shown, describes the variability, and presents the results of the statistical analysis just to back up these results. Here, this is probably best done within the text of your results. For the bull elephants in Example 4.1 you might say

> The weight of the bull elephants, with a mean and standard error of 4.70 ± 0.056, was 4% greater than the expected weight of 4.5 tonnes, a difference that a one-sample *t* test showed was significant ($t_{15} = 3.55$, $P = 0.003$). Note that the subscript after *t* is the number of degrees of freedom.

4.7 The paired *t* test

4.7.1 Purpose

To test whether the means of two sets of **paired measurements** are different from each other. Examples might be a single group of people measured twice, e.g. before vs after a treatment; or related people measured once, e.g. older vs younger identical twins.

4.7.2 Rationale

The idea behind the paired *t* test is that you look at the difference between each pair of points and then see whether the mean of these values is different from 0. The test therefore has two stages. You first calculate the difference d between each of the paired measurements you have made. You can then use these figures to calculate the mean difference between the two sets of measurements and the standard error of the difference. You then use a one-sample *t* test to determine whether the mean difference $\bar{d}$ is different from zero. The test statistic *t* is the number of standard errors the difference is away from zero. It can be calculated using a calculator or using computer programs using the following equation.

$$t = \frac{\text{Mean difference}}{\text{Standard error of difference}} = \frac{\bar{d}}{SE_d} \tag{4.3}$$

This procedure has the advantage that it removes the variability within each sample, concentrating only on the *differences* between each pair, so it improves your chances of detecting an effect.

4.7.3 Validity

Both sets of data must be normally distributed.

4.7.4 Carrying out the test

To see how to carry out a test in practice it is perhaps best to run through a straightforward example.

Example 4.2

Two series of measurements were made of the pH of nine ponds: at dawn and at dusk. The results are shown below.

Pond	Dawn pH	Dusk pH	Difference
1	4.84	4.91	0.07
2	5.26	5.62	0.36
3	5.03	5.19	0.16
4	5.67	5.89	0.22
5	5.15	5.44	0.29
6	5.54	5.49	−0.05
7	6.01	6.12	0.11
8	5.32	5.61	0.29
9	5.44	5.70	0.26
$\bar{x}$	5.362	5.552	0.190
s	0.352	0.358	0.129
SE	0.1174	0.1194	0.0431

Carrying out descriptive statistics shows that the mean difference $\bar{d} = 0.19$ and the standard error of the difference $\overline{SE}_d = 0.043$.

Do the ponds have a different pH at these times?

Solution

Step 1: Formulating the null hypothesis

The null hypothesis is that the mean difference $\bar{d}$ *is not* different from zero.

Here, the null hypothesis is that the mean of the differences in the pH *is* 0, i.e. the ponds have the same pH at dawn and dusk.

Step 2: Calculating the test statistic

Using a calculator

Using equation 3.3, we can calculate that $t = 0.190/0.0431 = 4.40$. The difference is 4.40 standard errors away from zero.

Using SPSS

SPSS can readily work out *t* as well as other important elements in the test. Simply put the data side by side into two columns so that each pair of observations has a single row and give each column a name (here dawnph and duskph). Next, click on the **Analyze** menu, move onto the **Compare Means** bar, and click on the **Paired-Samples T Test** bar. SPSS will produce the **Paired-Samples T Test** dialogue box. Put the columns to be compared into the **Variables** box as shown below by clicking on both of them and then on the arrow. The completed data and dialogue box are shown below.

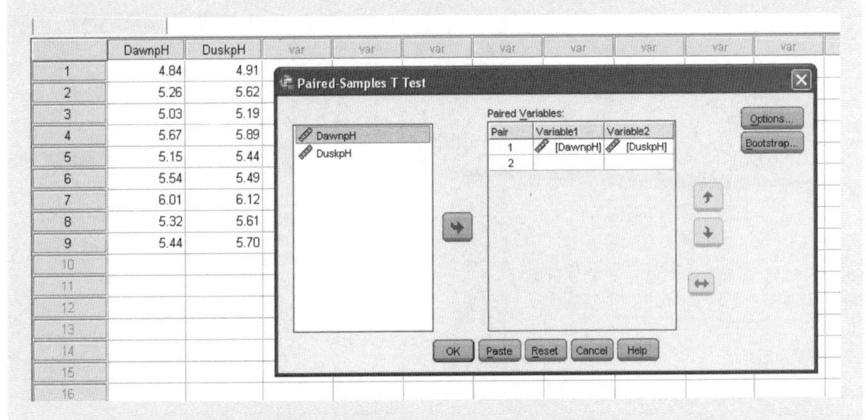

Finally click on **OK** to run the test.

SPSS will produce the following relevant output:

Paired sample statistics

		Mean	*N*	Std. deviation	Std. error mean
Pair 1	DawnpH	5.3622	9	0.35220	0.11740
	DawkpH	5.5522	9	0.35818	0.11939

Paired samples test

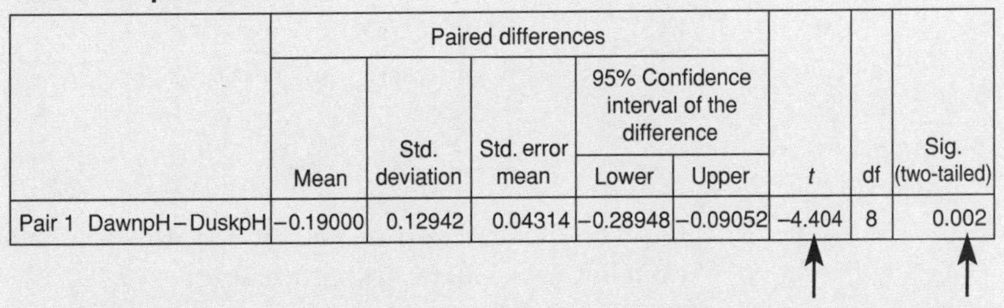

	Paired differences							
				95% Confidence interval of the difference				
	Mean	Std. deviation	Std. error mean	Lower	Upper	*t*	df	Sig. (two-tailed)
Pair 1 DawnpH−DuskpH	−0.19000	0.12942	0.04314	−0.28948	−0.09052	−4.404	8	0.002

SPSS gives the descriptive statistics in the first table and in the last table calculates *t*, which here is −4.404. (Note that version 19 of SPSS always takes the column which is second in the alphabet from that which is first.)

Using MINITAB

Simply put the data side by side into two columns so that each pair of observations has a single row and give each column a name (here Dawn pH and Dusk pH). Next, click on the **Stat** menu, move onto the **Basic Statistics** bar, and click on **Paired t**. MINITAB will produce the **Paired t (Test and Confidence Interval)** dialogue box. Put the columns to be compared into the **First Sample** and **Second Sample** box as shown. The completed data and dialogue box are shown below.

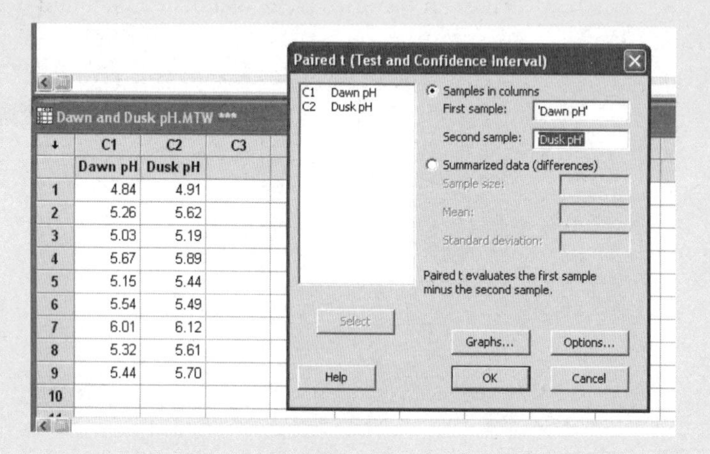

↓	C1	C2	C3
	Dawn pH	Dusk pH	
1	4.84	4.91	
2	5.26	5.62	
3	5.03	5.19	
4	5.67	5.89	
5	5.15	5.44	
6	5.54	5.49	
7	6.01	6.12	
8	5.32	5.61	
9	5.44	5.70	
10			

Finally click on **OK** to run the test.

MINITAB will produce the following relevant output:

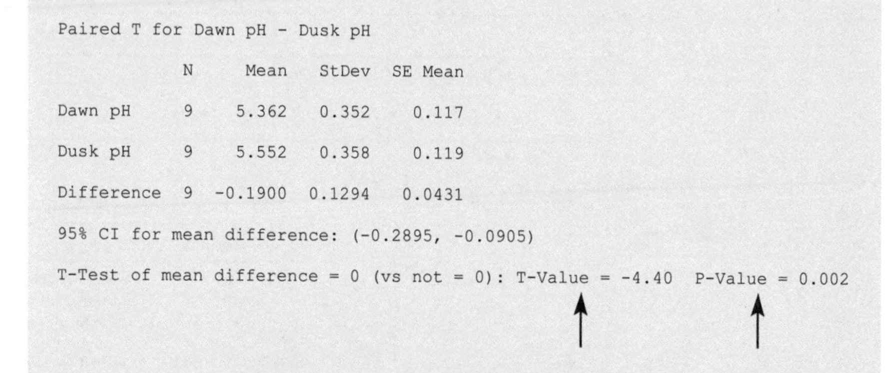

```
Paired T for Dawn pH - Dusk pH

             N     Mean    StDev  SE Mean

Dawn pH      9    5.362    0.352    0.117

Dusk pH      9    5.552    0.358    0.119

Difference   9  -0.1900   0.1294   0.0431

95% CI for mean difference: (-0.2895, -0.0905)

T-Test of mean difference = 0 (vs not = 0): T-Value = -4.40  P-Value = 0.002
```

It gives the descriptive statistics in the top table and the value of t (T-value) as -4.40 in the bottom line

Step 3: Calculating the significance probability

You must calculate the probability P that the absolute value of the test statistic would be equal to or greater than t if the null hypothesis were true.

Using a calculator

You must compare your value of $| t |$ with the critical value of the t statistic for $(N - 1)$ degrees of freedom where N is the number of pairs of observations, and at the 5% level ($t_{(N-1)}(5\%)$). This is given in Table S1 at the end of the book.

Here there are $9 - 1 = 8$ degrees of freedom, so the critical value of t for the 5% level is 2.306.

Using SPSS and MINITAB

SPSS and MINITAB directly work out the probability. P, and calls it Sig. (two-tailed) or P-value.

Here Sig. (two-tailed) = P-value = 0.002.

Step 4: Deciding whether to reject the null hypothesis

Using a calculator

- If $|t|$ is greater than or equal to the critical value, you must reject the null hypothesis. Therefore you can say that the mean difference is significantly different from zero.
- If $|t|$ is less than the critical value, you have no evidence to reject the null hypothesis. Therefore you can say that the mean difference is not significantly different from zero.

Here $| t | = 4.40 > 2.306$.

Using SPSS and MINITAB

- If Sig. (two-tailed) or P-value ≤ 0.05 you must reject the null hypothesis. Therefore you can say that the mean difference is significantly different from zero.
- If Sig. (two-tailed) or P-value > 0.05 you have no evidence to reject the null hypothesis. Therefore you can say that the mean difference is not significantly different from zero.

Here Sig. (two-tailed) = P-value = $0.002 < 0.05$.

Therefore we must reject the null hypothesis. We can say that the mean difference between dawn and dusk is significantly different from zero. In other words, the pH of ponds is significantly different at dusk from at dawn; in fact it's higher at dusk.

Step 5: Calculating confidence limits

The 95% confidence limits for the mean difference are given by the equation

$$95\%\ \text{CI(difference)} = \bar{d} \pm (t_{(N-1)}(5\%) \times \overline{SE}_d) \tag{4.4}$$

Using a calculator

The 95% confidence intervals can be calculated from the equation

$$\begin{aligned} 95\%\ \text{CI(difference)} &= 0.19 \pm (2.306 \times 0.043) \\ &= 0.09 \text{ to } 0.29. \end{aligned}$$

It is 95% likely that the pH at dusk will be between 0.09 and 0.29 higher than the pH at dawn.

Using SPSS and MINITAB

SPSS and MINITAB give the 95% confidence interval for the difference as −0.0905 to −0.2895. Note that the difference is negative because they have calculated dawnph–duskph, not the reverse.

4.7.5 Presenting the results

You could present the results of your paired t test in the text or in a table, but probably the best way is to present the means and standard errors of the two sets of data in a bar chart. You can then refer to this chart in the text, where you present the results of the t test itself. For the results of the pH of the ponds you would present the results as follows.

The mean pH of the ponds at dawn and dusk is shown in Figure 4.2. A paired t test showed that the pH was significantly higher at dusk than dawn ($t_8 = 4.40$, $P = 0.002$).

Note that a paired t test can be significant even if the error bars on your bar chart overlap each other. This is because the paired t test removes the within-sample variability.

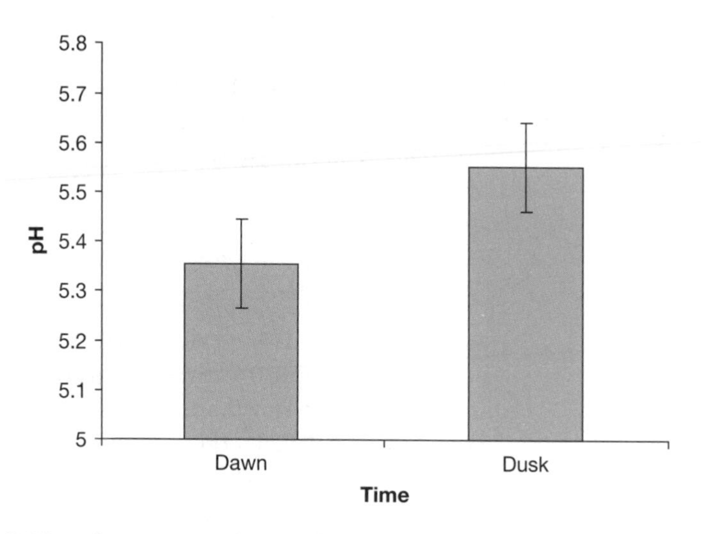

Figure 4.2 Mean (± standard error) of the pH of the nine ponds at dawn and dusk.

4.8 The two-sample *t* test

4.8.1 Purpose

To test whether the means of a two sets of **unpaired** measurements are different from each other. For instance it is used to test whether experimentally treated organisms are different from controls, or one species is different from another.

4.8.2 Rationale

standard error of the difference ($\overline{SE}_d$)
A measure of the spread of the difference between two estimated means.

This test is rather more complex than the previous two because you have to decide the probability of overlap between the distributions of *two* sample means (Figure 4.3). To do this you have to calculate *t* by comparing the difference in the means of the two populations with an estimate of the **standard error of the difference** between the two populations, using the equation

$$t = \frac{\text{Mean difference}}{\text{Standard error of difference}} = \frac{\bar{x}_A - \bar{x}_B}{\overline{SE}_d} \tag{4.5}$$

In this case, however, it is much more complex to calculate the standard error of the difference $\overline{SE}_d$ because this would involve comparing each member of the first population with each member of the second. Using a calculator $\overline{SE}_d$ can be estimated if we assume that the variance of the two populations is the same. It is given by the equation

$$\overline{SE}_d = \sqrt{(\overline{SE}_A)^2 + (\overline{SE}_B)^2} \tag{4.6}$$

where $\overline{SE}_A$ and $\overline{SE}_B$ are the standard errors of the two populations. If the populations are of similar size, $\overline{SE}_d$ will be about one-and-a-half times as big as either population standard error. Computer packages can also perform a more complex calculation of $\overline{SE}_d$ that makes no such simplifying assumption.

4.8.3 Validity

Both sets of data must be normally distributed. The two-sample *t* test also makes an important assumption about the measurements: it assumes the two sets of measurements are **independent** of each other. This would not be true of the data on the ponds we examined in Example 4.2, because each measurement has a pair, a measurement from the same pond at a different time of day. Therefore it is not valid to carry out a two-sample *t* test on this data.

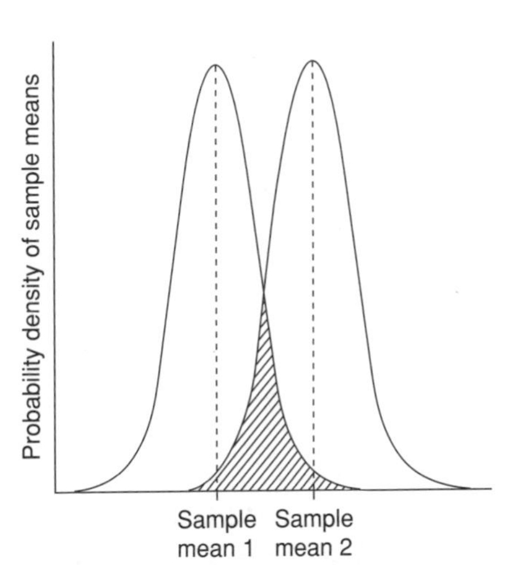

Figure 4.3 Overlapping populations. The estimated probability distributions of two overlapping populations worked out from the results of samples.

4.8.4 Carrying out the test

To see how to carry out a test in practice it is perhaps best to run through a straightforward example.

Example 4.3

The following data were obtained by weighing 16 cow elephants as well as the 16 bull elephants we have already weighed. We will test whether bull elephants have a different mean mass from cow elephants.

Masses of bull elephants (tonnes)

4.6	5.0	4.7	4.3	4.6	4.9	4.5	4.6
4.8	4.5	5.2	4.5	4.9	4.6	4.7	4.8

Masses of cow elephants (tonnes)

4.3	4.6	4.5	4.4	4.7	4.1	4.5	4.4
4.2	4.3	4.5	4.4	4.5	4.4	4.3	4.3

Solution

Carrying out descriptive statistics yields the following results:

Bull elephants: mean = 4.70, $s = 0.23$, $\overline{SE} = 0.056$

Cow elephants: mean = 4.40, $s = 0.15$, $\overline{SE} = 0.038$

It looks like bulls are heavier, but are they significantly heavier?

Step 1: Formulating the null hypothesis

The null hypothesis is that the mean of the differences *is not* different from zero. In other words, the two groups have the same mean.

Here the null hypothesis is that the bull and cow elephants have the same mean weight.

Step 2: Calculating the test statistic

Using a calculator

Using equations 3.5 and 3.6 we can calculate that

$$t = 4.70 - 4.40/\sqrt{(0.056^2 + 0.038^2)}$$
$$= 0.30/0.068 = 4.43.$$

The means are 4.43 pooled standard errors apart.

Using SPSS

To perform a two-sample t test in SPSS, you must first *put all the data into the same column* because each measurement is on a different organism. Call it something like **weight**. To distinguish between the two groups, you must create a second, subscript, column with one of two values, here 1 and 2. We will call it **sex**. You can also identify the subscripts using the **Values** box in the **Variable View** screen. To do this click on the box and onto the three dots on the right. The **Value Labels** dialogue box will come up. Type the first subscript (here 1) into the **Value** box and type in its name (here bulls) in the **Label** box. Click on **Add** to save this, then do the same for 2 and cows. The completed box is shown below.

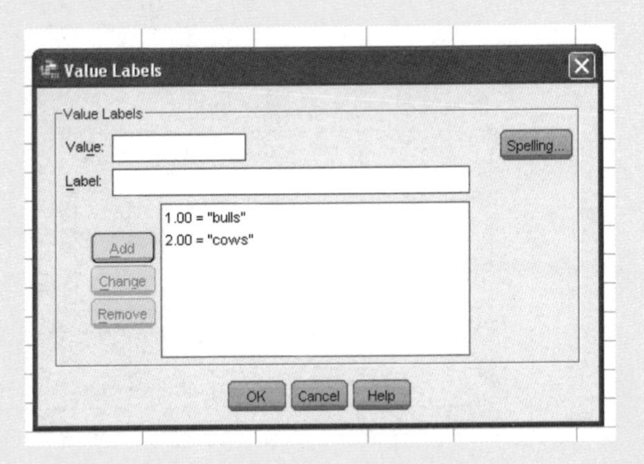

Once you have entered your data, simply click on the **Analyze** menu, move onto the **Compare Means** bar, and click on **Independent-Samples T Test.** SPSS will come up with the **Independent-Samples T Test** dialogue box.

Put the variable you want to test (here **weight**) into the **Test Variable** box and the subscript column (here **sex**) into the **Grouping Variable** box. Define the groups by clicking on the **Define Groups** tab to bring up the **Define Groups** dialogue box. Put in the values of the subscript column (here 1 and 2) into that box to give the data set and completed boxes shown below.

	weight	sex
1	4.60	1.00
2	5.00	1.00
3	4.70	1.00
4	4.30	1.00
5	4.60	1.00
6	4.90	1.00
7	4.50	1.00
8	4.60	1.00
9	4.80	1.00
10	4.50	1.00
11	5.20	1.00
12	4.50	1.00
13	4.90	1.00
14	4.60	1.00
15	4.70	1.00
16	4.80	1.00
17	4.30	2.00
18	4.60	2.00
19	4.50	2.00
20	4.40	2.00
21	4.70	2.00
22	4.10	2.00
23	4.50	2.00

Click on **Continue** to get back to the main dialogue box and finally click on **OK** to run the tests. SPSS comes up with the following results.

Group statistics

	sex	*N*	Mean	Std. deviation	Std. error mean
weight	bull	16	4.7000	0.22509	0.05627
	cow	16	4.4000	0.15055	0.03764

Independent samples test

		Levene's test for equality of variances		*t* test for equality of means						
									95% Confidence interval of the difference	
		F	Sig.	*t*	df	Sig. (2-tailed)	Mean difference	Std. error difference	Lower	Upper
Weight	Equal variances assumed	2.301	0.140	4.431	30	0.000	0.30000	0.06770	0.16174	0.43826
	Equal variances not assumed			4.431	26.183	0.000	0.30000	0.06770	0.16089	0.43911

In the first box it gives the descriptive statistics for the two sexes (note that I have given names for the values of the subscripts 1 and 2) and then performs the *t* test, both with and without making the assumption of equal variances. In both cases here, $t = 4.431$. They are usually similar. The first test is only valid if the variances are not significantly different, and this is, in fact, tested by Levene's test for equality of variances, the results of which are given at the left of the second table. If Sig. < 0.05 then the test is not valid. Here Sig. $= 0.140$, so you could use either test. If in doubt though, use the second, more accurate statistic.

Using MINITAB

There are two ways in which you can input your data into MINITAB to perform a two-sample *t* test. You can either put the results from the two samples into separate columns, or (I recommend this so that later more complex tests don't come as so much of a shock) put all the data into the same column as in SPSS. Call it something like **weight**. To distinguish between the two groups, you must create a second, subscript, column with one of two values, here 1 and 2. We will call it **sex**.

Once you have entered your data, simply click on the **Stat** menu, move onto the **Basic Statistics** bar, and click on **2-Sample t.** MINITAB will come up with the **2-Samples t (Test and Confidence Interval)** dialogue box.

Put the variable you want to test (here **weight**) into the **Samples** box and the subscript column (here **sex**) into the **Subscripts** box to give the data set and completed boxes shown below.

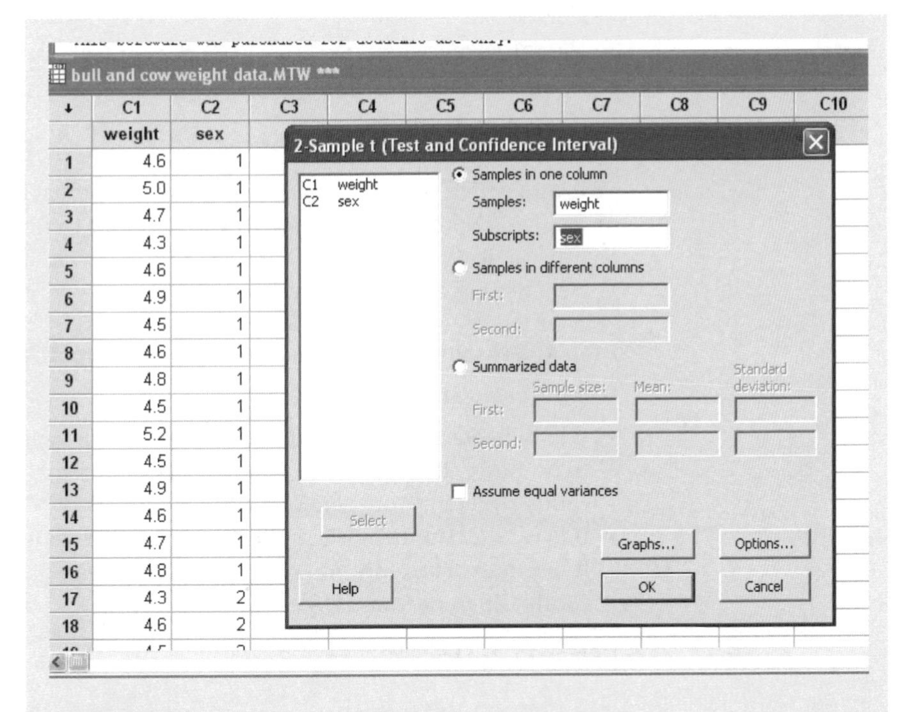

Click on **OK** to run the tests.

MINITAB comes up with the following results.

```
Two-sample T for weight

sex   N   Mean   StDev  SE Mean

1    16  4.700  0.225    0.056

2    16  4.400  0.151    0.038

Difference = mu (1) - mu (2)

Estimate for difference:  0.3000

95% CI for difference:  (0.1608, 0.4392)

T-Test of difference = 0 (vs not =): T-Value = 4.43  P-Value = 0.000  DF =

26
```

The means, standard deviations and standard errors are given at the top and the value of t (*T*-value) on the final line. Here *T*-value = 4.43.

Step 3: Calculating the significance probability

Using a calculator

You must compare your value of $|t|$ with the critical value of the t statistic for $N_A + N_B$ − two degrees of freedom, where N_A and N_B are the sample sizes of groups A and B.

Here there are 16 + 16 − 2 = 30 degrees of freedom; the critical value of t for the 5% level is 2.042.

Using SPSS and MINITAB

SPSS and MINITAB will directly work out the probability, Sig. (two-tailed) or *P*-value. (Note that the bigger the value of $|t|$, the smaller the value of Sig. (two-tailed) or *P*-value).

Here Sig. (two-tailed) = *P*-value = 0.000.

Step 4: Deciding whether to reject the null hypothesis

Using a calculator

- If $|t|$ is greater than or equal to the critical value, you must reject the null hypothesis. Therefore you can say that the mean difference is significantly different from zero.
- If $|t|$ is less than the critical value, you have no evidence to reject the null hypothesis. Therefore you can say that the mean difference is not significantly different from zero.

Here 4.43 > 2.042.

Using SPSS and MINITAB

- If Sig. (two-tailed) or *P*-value ≤ 0.05 you must reject the null hypothesis. Therefore you can say that the mean difference is significantly different from zero.
- If Sig. (two-tailed) or *P*-value > 0.05 you have no evidence to reject the null hypothesis. Therefore you can say that the mean is not significantly different from zero.

Here Sig. (two-tailed) = *P*-value = 0.000 < 0.05

Therefore we must reject the null hypothesis. We can say that the mean weights of the bull and cow elephants was different. In fact the bulls were significantly heavier than the cows.

Step 5: Calculating confidence limits

The 95% confidence intervals for the mean difference are given by the equation

$$95\%\ \text{CI(difference)} = \bar{x}_A - \bar{x}_B \pm (t_{N_A+N_B-2}(5\%) \times \overline{SE}_d) \qquad (4.7)$$

Using a calculator

$$95\%\ \text{CI(difference)} = 4.70 - 4.40 \pm (2.042 \times 0.0680$$
$$= 0.16 \text{ to } 0.44$$

Using SPSS and MINITAB

SPSS and MINITAB calculate these limits directly and you should round up to two significant figures to give 0.16 to 0.44.

4.8.5 Presenting the results

If you have investigated just one or two measurements probably the best way to present your data is in a bar chart with the means and standard errors of the two sets of data. You can then refer to this chart in the text, where you can emphasise the size of the difference and back this up with the results of the t test itself. For the results of the bull and cow elephants you would present the results as follows.

The mean weights of the bull and cow elephants is shown in Figure 4.4. The bull elephants were on average *x*% heavier than the cows, a difference that a two-sample t test showed was significant 6.8 ($t_{30} = 4.43$, $P < 0.0005$).

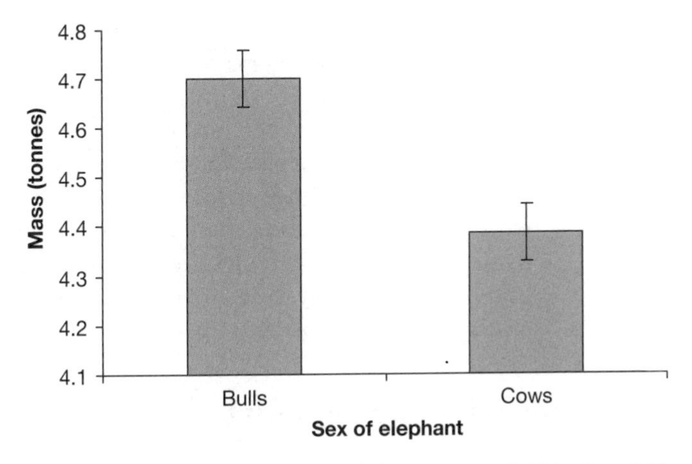

Figure 4.4 The mean (± standard error) of the masses of 16 bull and 16 cow elephants.

Note that the P value given is not zero as SPSS and MINITAB report. Indeed P values are *never* zero, so when a computer package gives a value of 0.000 it just means that it is less than 0.0005. A two-sample t test is usually only significant if the error bars on your bar chart do not overlap each other.

If you have taken several measurements on the same samples (for instance comparing a control and experimentally treated group) it is usually best to present the information in the form of a table as in Table 4.1, with the benefit of an informative legend.

Table 4.1 The effect of nitrogen treatment on sunflower plants. The results show the means ±standard error for control and high nitrogen plants of their height, biomass, stem diameter and leaf area.

Treatment	n	Height (cm)	Biomass (g)	Stem diameter (mm)	Leaf area (cm^2)
Control	12	36.5 ± 3.6	14.3 ± 2.8	9.6 ± 1.9	194 ± 38
High N	12	50.4 ± 3.9 **	22.4 5 ± 3.2 ***	10.9 ± 2.4 NS	286 ± 32*

Asterisks denote the degree of significance: * $P < 0.05$; ** $P < 0.01$; *** $P < 0.001$; NS no significant difference.

The use of asterisks to denote the degree of significance is a common method and one that is very concise and clear. Having produced the table, you should refer to it in the text of the results section, saying something like

The effects of supplementary nitrogen are summarised in Table 4.1. It can be seen that supplementary nitrogen increased height by around 38%, biomass by 57% and leaf area by 46%, but had no significant effect on stem diameter.

4.9 Introduction to non-parametric tests for differences

rank
Numerical order of a data point.

median
The central value of a distribution (or average of the middle points if the sample size is even).

Non-parametric tests make no assumption about the shape of the distribution but use only information about the **rank** of each data point. The tests for differences compare the **medians** of the groups instead of comparing means, and all the tests look at the probability of getting the ranked data points in a particular order. This means that the tests are intuitively fairly easy to understand. However, it often takes a great deal of time to assign ranks, and to manipulate these ranks to produce the test statistic. Therefore it is often quicker to carry out non-parametric statistics using computer packages which do all that for you. For this reason we will take a brief look at the rationale and mathematics of each of the non-paremetric equivalents of *t* tests, before looking at how to carry it out both on a calculator and using statistical packages.

4.10 The one-sample sign test

4.10.1 Purpose

To test whether the sample median of one measurement taken on a single population is different from an expected value *E*. It is the non-parametric equivalent of the one-sample *t* test.

4.10.2 Rationale

Like the one-sample *t* test, the first stage of the one sample sign test is to calculate the difference *d* between each measurement you have made and the expected value, *E*. Next, you rank the absolute values of these differences, and give the positive differences a plus and the negative differences a minus sign. Finally, you add all the negative ranks together, and separately add all the positive ranks together to give two values of T: $T-$ and $T+$. Note that if the median of your sample is lower or higher than the expected median, one value of T will be large and the other will be small. In this test the smaller T value is compared with the critical values table for the relevant group size. The null hypothesis is rejected if your smaller value of T is **lower than or equal to** a critical value. Note that this test is a special case of the Wilcoxon matched pairs test (Section 4.11) but substituting expected values for one of the two samples.

4.10.3 Carrying out the test

The method of carrying out the test is best seen by working through an example.

Example 4.4

After a course on statistics, students were required to give their verdict on the merits of the course in a questionnaire. Students could give scores of 1 = rubbish through 3 = reasonable to 5 = excellent. The following scores were obtained.

Number	
Students scoring 1	8
Students scoring 2	14
Students scoring 3	13
Students scoring 4	4
Students scoring 5	0

Did the course score significantly lower than 3?

Solution

Step 1: Formulating the null hypothesis

The null hypothesis is that the median of the scores the students gave was 3.

Step 2: Calculating the test statistic

Using a calculator

	N	(Score – *E*)	Ranks of (Score – *E*)
Students scoring 1	8	−2	22.5
Students scoring 2	14	−1	9.5
Students scoring 3	13	0	0
Students scoring 4	4	+1	9.5
Students scoring 5	0	+2	22.5

Sum of negative ranks: $T- = (14 \times 9.5) + (8 \times 22.5) = 133 + 180 = 313$
Sum of positive ranks: $T+ = (4 \times 9.5) = 38$
The smaller of the two ranks is the test statistic: $T = 38$.

The ties (where the scores equal the expected value and so Score $- E = 0$) contribute nothing to the comparison and are ignored. The ranks of the differences are given from 1 = the lowest to 26 = the highest. When there are ties between the ranks of the differences, each one is given the mean of the ranks. Hence in this case there are 18 students with a difference from expected of 1 (14 with −1 and 4 with +1). The mean of ranks 1 to 18 is 9.5, so each of these students is given 9.5. There are also eight students with a difference from the expected value of 2. The mean of ranks 19 to 26 is 22.5, so each of these students is given a rank of 22.5. Adding up the ranks gives a value for $T-$ of 313 and for $T+$ of 38. The smaller value is $T+$, so the T value to use is 38.

Using SPSS

Enter your data values into one column (here named, say, **scores**) and the expected mean values (here 3) as a second column (named, say **expected**). Next, click on the **Analyze** menu, move onto the **Nonparametric Tests** bar, onto **Legacy Dialogs** and click on the **two Related Samples** bar. SPSS will produce the **Two Related Samples Tests** dialogue box. Put the columns to be compared into the **Test Pair(s) List** box, making sure the **Wilcoxon** test type is ticked. To get descriptive statistics and quartiles, go into **Options** and click on both **Descriptives** and **Quartiles.** The completed boxes and data are shown below.

Finally, click on **Continue** to get back to the main dialogue box and onto **OK** to run the test. SPSS will come up with the following results.

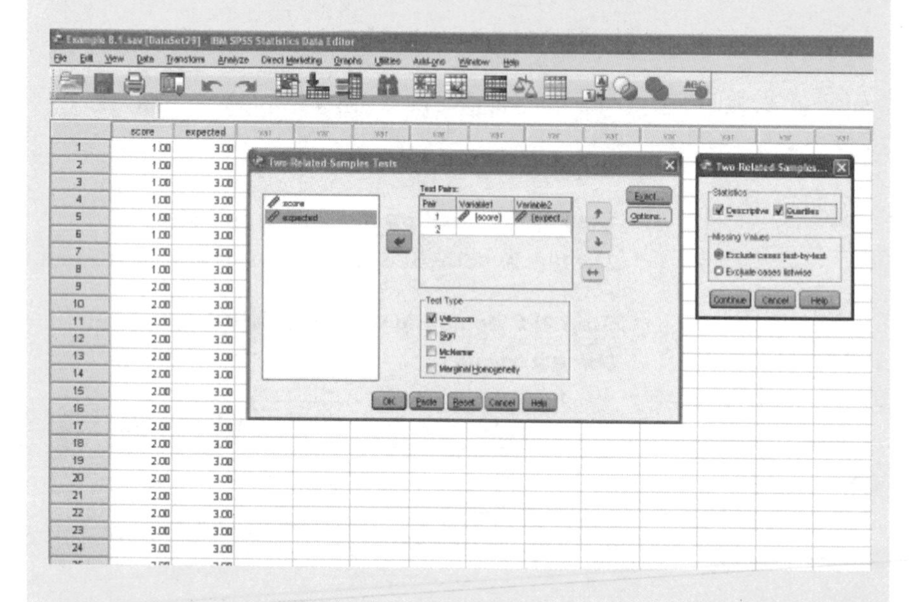

NPar tests

Descriptive statistics

	N	Mean	Std. deviation	Minimum	Maximum	Percentiles		
						25th	50th (Median)	75th
Score	39	2.3333	0.92717	1.00	4.00	2.0000	2.0000	3.0000
Expected	39	3.0000	0.00000	3.00	3.00	3.0000	3.0000	3.0000

Wilcoxon signed ranks test

Ranks

		N	Mean rank	Sum of ranks
Expected - score	Negative ranks	4[a]	9.50	38.00
	Positive ranks	22[b]	14.23	313.00
	Ties	13[c]		
	Total	39		

[a] Expected < score.

[b] Expected > score.

[c] Expected = score.

Test statistics[b]

	Expected-score
Z	−3.651[a]
Asymp. Sig. (two-tailed)	0.000

[a] Based on negative ranks.

[b] Wilcoxon signed ranks test.

As well as the descriptive statistics SPSS has given the sums of the ranks, $T- = 38$ and $T+ = 313$, and has also given something called the Z statistic. Here $Z = -3.651$.

NB. SPSS now has new analysis available for this test, available through the **One Sample** bar. I don't recommend this new method, however, as though it carries out the test it doesn't actually present you with any statistics!

Using MINITAB

Enter your data values into one column (here named, say, **scores**). Next, click on the **Stat** menu, move onto the **Nonparametrics** bar, then click on the **1-Sample Sign** bar. MINITAB will produce the **1-Sample Sign** dialogue box. Put the column to be compared into the **Variables** box, tick on **Test Median** and enter the value you are testing against (here 3). The completed box and data are shown below.

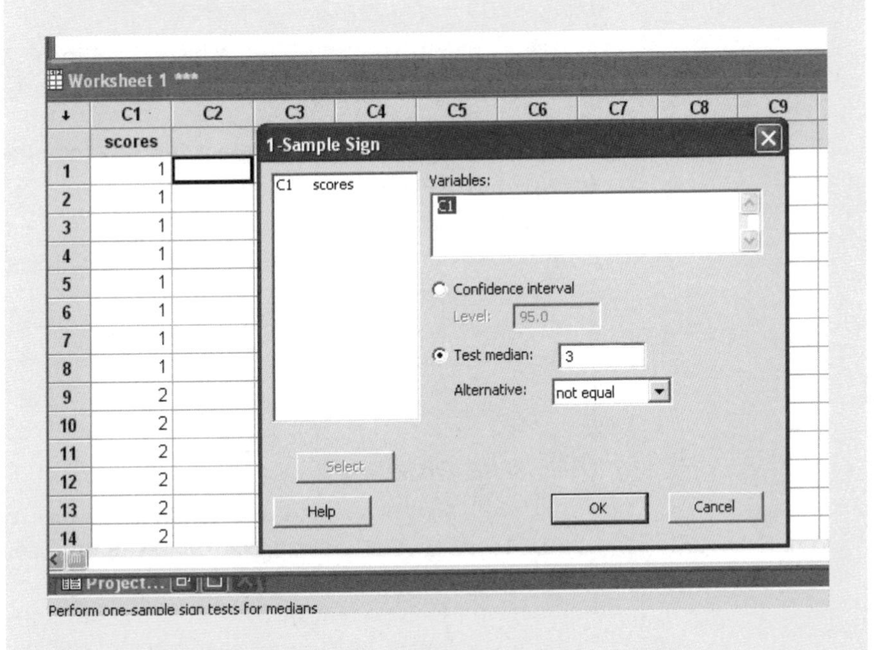

Finally, click on **OK** to run the test. MINITAB will come up with the following results.

Sign Test for Median: scores

```
Sign test of median =  3.000 versus not = 3.000

          N  Below  Equal  Above        P  Median

scores   39     22     13      4  0.0005   2.000
                                    ↑
```

MINITAB fails to give any actual statistics!

Step 3: Calculating the significance probability

You must calculate the probability P that the test statistic T would be **less than or equal to** T if the null hypothesis were true.

Using a calculator

You must compare your value of $|t|$ with the critical value of the Wilcoxon T distribution for $N - 1$ degrees of freedom, where N is the number of non-tied pairs. This is given in Table S4 at the end of the book.

Here there are 39 students, but 13 have the expected median score, so are tied, and hence $N = 39 - 13 = 26$. The critical value of T for $26 - 1 = 25$ degrees of freedom at the 5% level is 89.

Using SPSS and MINITAB

SPSS and MINITAB both show that the probability of getting such test statistics is low (SPSS rounds Sig. to 0.000, MINITAB rounds P to 0.0005).

Step 4: Deciding whether to reject the null hypothesis

Using a calculator

- If $|t|$ is **less than** or equal to the critical value, you must reject the null hypothesis. You can say that the distribution is significantly different from expected.
- If $|t|$ is greater than the critical value, you cannot reject the null hypothesis. You have found no significant difference from the expected distribution.

Here $T_{25} = 38 < 89$.

Using SPSS and MINITAB

- If Asymp. Sig. (two-tailed) or $P \leq 0.05$ you must reject the null hypothesis. Therefore you can say that the mean difference is significantly different from zero.
- If Asymp. Sig. (two-tailed) or $P > 0.05$ you have no evidence to reject the null hypothesis. Therefore you can say that the mean is not significantly different from zero.

Here Asymp. Sig. (two-tailed) = 0.000: $P = 0.0005 < 0.05$.

Therefore we must reject the null hypothesis. We can say that the median scores of the course were significantly different from the expected value of 3. In fact the median score (2) was lower, showing the unpopularity of this course.

4.10.4 Presenting the results

The results of a sign test are best presented in text. For the test example you might say

> The median scores of the students was 2, which a one sample sign test showed was significantly lower than the expected value of 3 ($T_{25} = 38$, $P < 0.0005$).

4.11 The Wilcoxon matched pairs test

4.11.1 Purpose

To test whether the **medians** of **two paired** measurements made on identifiable population are different from each other. Examples might be a single group of people measured twice, e.g. before and after a treatment; or related people measured once, e.g. a single group of husbands and wives. This in the non-parametric equivalent of the paired *t* test.

4.11.2 Rationale

Like the paired *t* test, the first stage of the Wilcoxon test is to calculate the difference d between each of the two paired measurements you have made. Next, you rank the absolute value of these differences, and give the positive differences a plus and the negative differences a minus sign. Finally, you add all the negative ranks together, and separately add all the positive ranks together to give two values of T: $T-$ and $T+$. Note that if one set of measurements is larger than the other, one value of T will be large and the other will be small. In this test the smaller T value is compared with the critical values table for the relevant group sizes. The null hypothesis is rejected if your smaller value of T is **lower than or equal to** a critical value.

4.11.3 Carrying out the test

The method of carrying out the test is best seen by working through an example.

Example 4.5

A new treatment for acne was being tested. Ten teenage sufferers had their level of acne judged on an arbitrary rank scale from 0 (totally clear skin) through to 6 (very bad acne). They were then given the new treatment for 4 weeks and assessed again. The following results were found:

Level before treatment	Level after treatment
4	3
5	2
3	1
4	5
5	3
6	3
4	4
4	3
5	3
5	2

It looks as if the treatment reduces the severity of the acne, but is this a significant difference?

Solution

Step 1: Formulating the null hypothesis

The null hypothesis is that there is no difference in the median level before and after treatment.

Step 2: Calculating the test statistic

Using a calculator

You must first work out the differences, d, and ranks of the differences

Level before treatment	Level after treatment	d	Rank of d
4	3	−1	−2
5	2	−3	−8
3	1	−2	−5
4	5	+1	+2
5	3	−2	−5
6	3	−3	−8
4	4	0	0
4	3	−1	−2
5	3	−2	−5
5	2	−3	−8

The ties (where the level before equals the level after and so $d = 0$) contribute nothing to the comparison and are ignored. The ranks of the differences are given from 1 = the lowest to 9 = the highest. When there are ties between the ranks of the differences, each one is given the mean of the ranks. Hence in this case the three lowest differences are all equal to 1 (where d is either −1 or +1). They would be given ranks 1, 2 and 3, the mean of which is 2. Similarly there are three points that have the largest difference of 3; these points would have the ranks of 7, 8 and 9, the mean of which is 8.

The sum of minus ranks: $T- = (2 + 8 + 5 + 5 + 8 + 2 + 5 + 8) = 43$

The sum of plus ranks: $T+ = 2$

The smaller of the two ranks is the test statistic: $T = 2$.

Using SPSS

Enter the data into two columns, named, say, **skinbefore** and **skinafter**. Next, click on the **Analyze** menu, move onto the **Nonparametric tests** bar, onto **Legacy Dialogs** then click on the **Two Related Samples** bar. SPSS will produce the **Two Related Samples Test** dialogue box. Put the columns to be compared into the **Test Pair(s) List** box, making sure the **Wilcoxon** test type is ticked. To get descriptive statistics and quartiles, go into **Options** and click on both **Descriptives** and **Quartiles.** The completed boxes and data are shown below.

Finally, click on **Continue** to get back to the main dialogue box and then click on **OK** to run the test. SPSS will come up with the following results.

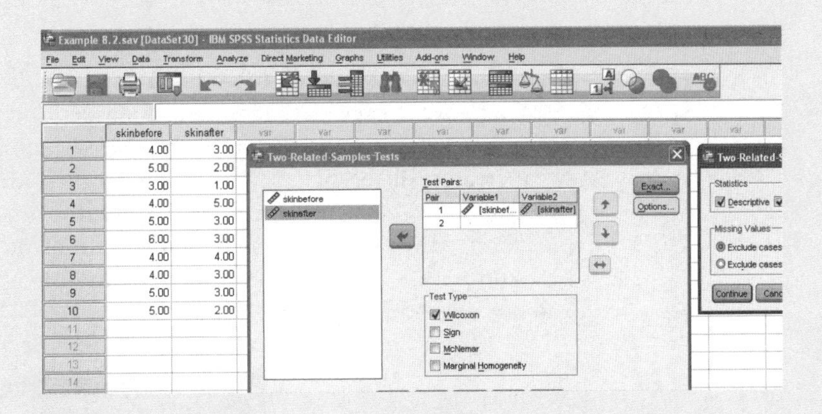

NPar tests

Descriptive statistics

	N	Mean	Std. deviation	Minimum	Maximum	Percentiles		
						25th	50th (Median)	75th
Skinbefore	9	4.5556	0.88192	3.00	6.00	4.0000	5.0000	5.0000
Skinafter	9	2.7778	1.09291	1.00	5.00	2.0000	3.0000	3.0000

Wilcoxon signed ranks test

Ranks

		N	Mean rank	Sum of ranks
Skinafter - skinbefore	Negative ranks	8[a]	5.38	43.00
	Positive ranks	1[b]	2.00	2.00 ←
	Ties	0[c]		
	Total	9		

Test statistics[b]

	Skinafter-skinbefore
Z	−2.455[a]
Asymp. Sig. (2-tailed)	0.014

[a] Based on positive ranks

[b] Wilcoxon signed ranks test

In the top table SPSS has given the sums of the ranks, $T- = 43$ and $T+ = 2$, and in the bottom one has also given something called the Z statistic. Here $Z = -2.455$.

NB SPSS now has new analysis available for this test, available through the **Related Samples** bar. I don't recommend this new method, however, as though it carries out the test it doesn't actually present you with any statistics!

Using MINITAB

To do this test in MINITAB is a bit awkward. You have to calculate the difference between each data item, and then perform a **1 Sample Sign Test.** Enter the data into two columns, named, say, **skinbefore** and **skinafter.** Next, create a **difference** column. Click on the **Calc** menu, and click onto the **Calculator** bar. This opens the **Calculator** diialogue box. Put the expression for the difference ('skin after' – 'skin before') into the **Expression** box, and put C3 into the **Store result in variable** box. The completed data and box are shown below.

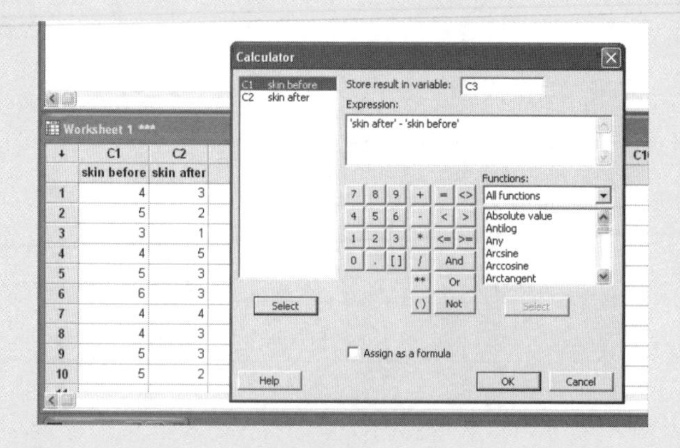

Click on **OK** to create the new column and name it **difference**. Finally, to run the test, click on the **Stat** menu, move onto the **Nonparametrics** bar, then click on the **1-Sample Wilcoxon** bar. MINITAB will produce the **1-Sample Wilcoxon** dialogue box. Put the **difference** into the **Variables** box, tick on **Test Median** and enter the value you are testing against (here 0). The completed box and data are shown below.

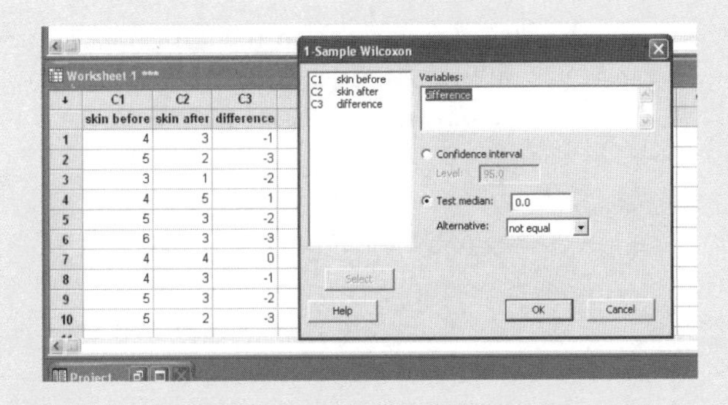

Finally, click on **OK** to run the test. MINITAB will come up with the following results.

Wilcoxon signed rank test: difference

```
Test of median = 0.000000 versus median not = 0.000000

                  N for   Wilcoxon              Estimated
              N    Test  Statistic        P        Median
difference   10       9        2.0    0.018        -1.500
```

MINITAB gives the Wilcoxon test statistic, here 2.0, in the table.

Step 3: Calculating the significance probability

You must calculate the probability P that the test statistic T would be **less than or equal to** T if the null hypothesis were true.

Using a calculator

You must compare your value of T with the critical value of the T statistic for N degrees of freedom, where N is the number of non-tied pairs. This is given in Table S4 at the end of the book.

Here there are 10 pairs, but there is one tie, so you must look up the critical T value for $10 - 1 = 9$ degrees of freedom. The critical value of T for the 5% level is 5.

Using SPSS and MINITAB

SPSS and MINITAB show that the probability Sig. or P of getting such test statistics is 0.014 and 0.018, respectively.

Step 4: Deciding whether to reject the null hypothesis

Using a calculator

- If T is **less than** or equal to than the critical value, you must reject the null hypothesis. Therefore you can say that the mean difference is significantly different from zero.

- If T is **greater than** the critical value, you have no evidence to reject the null hypothesis. Therefore you can say that the mean difference is not significantly different from zero.

  Here $T_9 = 2 < 5$.

Using SPSS and MINITAB

- If Asymp. Sig. (two-tailed) or $P \leq 0.05$ you must reject the null hypothesis. Therefore you can say that the mean difference is significantly different from zero.
- If Asymp. Sig. (two-tailed) or $P > 0.05$ you have no evidence to reject the null hypothesis. Therefore you can say that the mean is not significantly different from zero.

  Here Asymp. Sig. (two-tailed) = 0.014: $P = 0.018 < 0.05$.

Therefore we must reject the null hypothesis. We can say that the median acne scores were different after treatment from before. In fact the median score (3) was lower than before (5) so the treatment did seem to work.

4.11.4 Presenting the results

You can present the data using box and whisker plots as below.

The results of the Wilcoxon matched pairs test are then best presented in text. For the acne example you might say

The treatment improved the skin of the patients (Figure 4.5) reducing the median acne score from 5 to 3, a difference that a Wilcoxon matched pairs test showed was significant ($T_9 = 2$, $P = 0.018$).

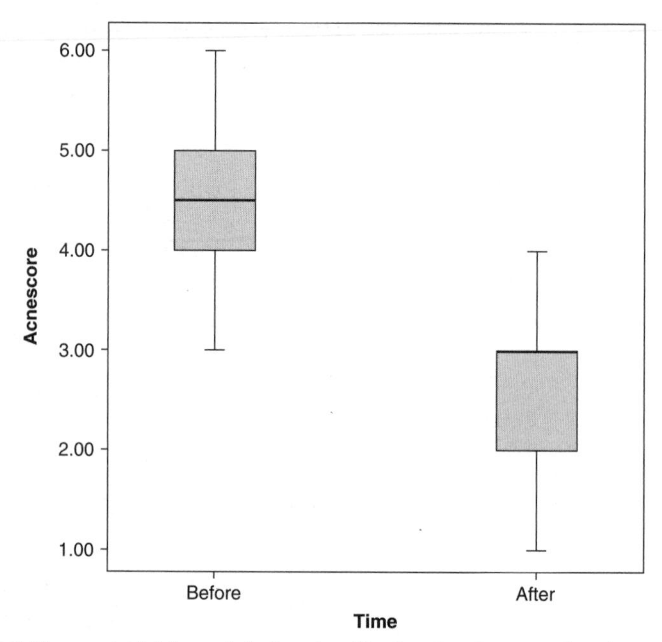

Figure 4.5 Box and whisker plot showing the levels of acne of patients before and after treatment.

4.12 The Mann–Whitney *U* test

4.12.1 Purpose

To test whether the medians of two **unpaired** sets of measurements are different from each other. For instance it is used to test whether experimentally treated organisms are different from controls, or one species is different from another.

4.12.2 Rationale

The Mann–Whitney U test works by comparing the ranks of observations in the two groups. First the data from the two groups are pooled and the ranks of each observation are calculated, with rank 1 being the smallest value. Where there are ties, observations are given the average value of the ranks. Next, the rank of each group is summed separately, to give the values R_1 and R_2. Finally two test statistics U_1 and U_2 are calculated using the following equations.

$$U_1 = n_1 n_2 + n_2(n_2 + 1)/2 - R_2 \tag{4.8}$$

$$U_2 = n_1 n_2 + n_1(n_1 + 1)/2 - R_1 \tag{4.9}$$

Where n_1 and n_2 are the sample sizes of group 1 and group 2, respectively. Note that if one group has much higher ranks than the other, one value of U will be large and the other will be small. In this test the *smaller* U value is compared with the critical values table for the relevant group sizes. Just like the Wilcoxon test, the null hypothesis is rejected if your value of U is **lower than or equal to** a critical value.

4.12.3 Carrying out the test

The method of carrying out the test is best seen by working through an example.

Example 4.6

In a field survey to compare the numbers of ground beetles in two fields, one of which had an arable crop, the other being permanent pasture, several pitfall traps were laid down and collected after a week.

The numbers of beetles caught in the traps in the two fields were as follows.

Field 1:	8,	12,	15,	21,	25,	44,	44,	60
Field 2:	2,	4,	5,	9,	12,	17,	19	

It looks as if more were caught in field 1, but is this a significant difference?

Solution

Step 1: Formulating the null hypothesis

The null hypothesis is that there is no difference in the median numbers of beetles between the two fields.

Step 2: Calculating the test statistic

Using a calculator

You must first work out the ranks of all the observations.

Observation:	2	4	5	8	9	12	12	15	17	19	21	25	44	44	60
Rank:	1	2	3	4	5	6.5	6.5	8	9	10	11	12	13.5	13.5	15

Field 2 observations are underlined.

Now $R_1 = 1 + 2 + 3 + 5 + 6.5 + 9 + 10 = 36.5$

$R_2 = 4 + 6.5 + 8 + 11 + 12 + 13.5 + 13.5 + 15 = 83.5$

Finally $U_1 = (7 \times 8) + 8(8 + 1)/2 - 83.5 = 8.5$

$U_2 = (7 \times 8) + 7(7 + 1)/2 - 36.5 = 47.5$

Take the lower value: $U = 8.5$

Using SPSS

As for the two-sample *t* test, you must first put all the data into the same column because each measurement is on a different trap. To distinguish between the two fields, you must create a second, subscript, column with one of two values, here 1 and 2. Now you can carry out the test. Simply click on the **Analyze** menu, move onto the **Nonparametric tests** bar, **Legacy Dialogs** and click on the **Two Independent Samples** bar. SPSS will come up with the **Two Independent Samples Tests** dialogue box. Put the variable you want to test (here **beetles**) into the **Test Variable** box, making sure the **Mann–Whitney** *U* test type is ticked. Put the subscript column (here **field**) into the **Grouping Variable** box. Define the groups by putting in the values of the subscript column (here 1 and 2). The completed boxes and data are shown below.

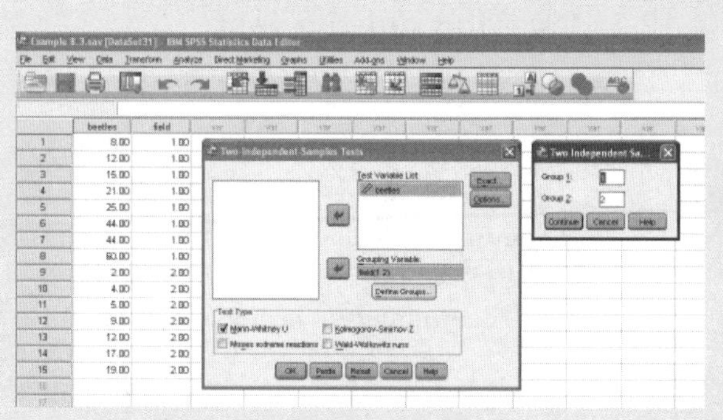

Finally click on **Continue** and then **OK** to run the test. SPSS will come up with the following tables.

Ranks

	Field	*N*	Mean rank	Sum of ranks
Beetles	1.00	8	10.44	83.50
	2.00	7	5.21	36.50
	Total	15		

Test statistics[b]

	Beetles
Mann–Whitney *U*	8.500
Wilcoxon *W*	36.500
Z	−2.261
Asymp. Sig. (2-tailed)	0.024
Exact Sig. [2*(1-tailed Sig.)]	021[a]

[a] Not corrected for ties

[b] Grouping variable: field

The Mann–Whitney *U* value, here 8.5, is given at the top of the last box.

NB SPSS now has new analysis available for this test, available through the **Independent Samples** bar. I don't recommend this new method, however, as though it carries out the test it doesn't actually present you with any statistics!

Using MINITAB

Unlike SPSS in MINITAB you have to put the data side by side into two separate columns, called something like **field one** and **field two**. Next, click on the **Stat** menu, move onto the **Nonparametrics** bar, and click on the **Mann–Whitney** bar. MINITAB will produce the **Mann–Whitney** dialogue box. Put the columns to be compared into the **First Sample** and **Second Sample** box as shown. The completed data and dialogue box are shown below.

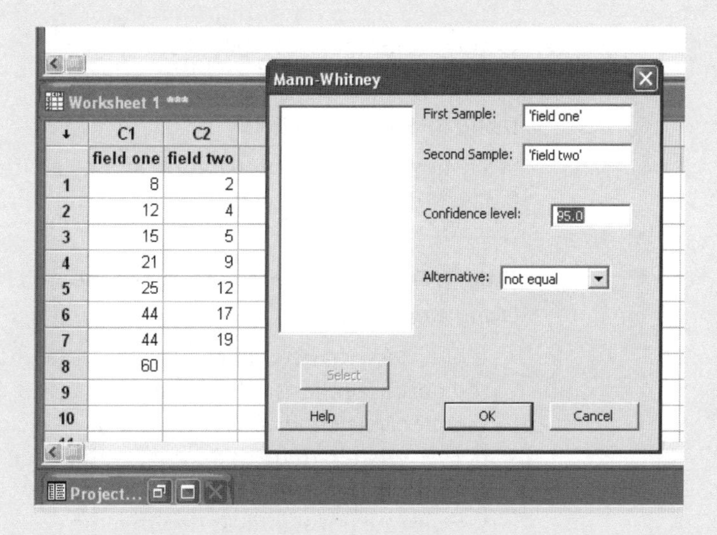

Finally, click on **OK** to run the test. MINITAB will come up with the following results. **Mann–Whitney Test and CI: field one, field two**

```
           N  Median
field one  8   23.00
field two  7    9.00
Point estimate for ETA1-ETA2 is 14.50
95.7 Percent CI for ETA1-ETA2 is (3.00,40.00)
W = 83.5
Test of ETA1 = ETA2 vs ETA1 not = ETA2 is significant at 0.0279
The test is significant at 0.0276 (adjusted for ties)
```

MINITAB gives the medians, but fails to give the U value.

Step 3: Calculating the significance probability

You must calculate the probability P that the test statistic U would be **less than or equal to** U if the null hypothesis were true.

Using a calculator

You must compare your value of U with the critical value of the U statistic for sample sizes of n_1 and n_2. This is given in Table S5 at the end of the book.

Looking up in the U distribution for $n_1 = 7$ and $n_2 = 8$ gives a critical value of U for the 5% level of 10.

Using SPSS and MINITAB

SPSS shows that the probability, Sig. of getting such a low value of U is 0.024. MINITAB gives significance probability as 0.0276.

Step 4: Deciding whether to reject the null hypothesis

Using a calculator

- If U is **less than or equal to** than the critical value, you must reject the null hypothesis. Therefore you can say that the medians of the two samples are significantly different from each other.
- If U is **greater than** the critical value, you have no evidence to reject the null hypothesis. Therefore you can say that the medians of the two samples are not significantly different.

Here $U_{7,8} = 8.5 < 10$.

Using SPSS and MINITAB

- If Asymp. Sig. (two-tailed) ≤ 0.05 you must reject the null hypothesis. Therefore you can say that the medians of the two samples are a significantly different from each other.
- If Asymp. Sig. (two-tailed) > 0.05, you have no evidence to reject the null hypothesis. Therefore you can say that the medians of the two samples are not significantly different.

Here Asymp. Sig. (two-tailed) = 0.024 or 0.0276 < 0.05.

Therefore we must reject the null hypothesis. We can say that the median numbers of beetles found in the two fields were different. In fact the median score (23) for field 1 was higher than that for field 2 (9).

4.12.4 Presenting the results

You can present the data using box and whisker plots. The results of the Mann–Whitney *U* test are then best presented in text. For the example of the beetles in the fields you might say

More beetles were found in field 1 than field 2 (Figure 4.6) with medians of 23 and 9 respectively. A Mann–Whitney *U* test showed that this difference was significant ($U_{7,8} = 8.5$, $P = 0.024$).

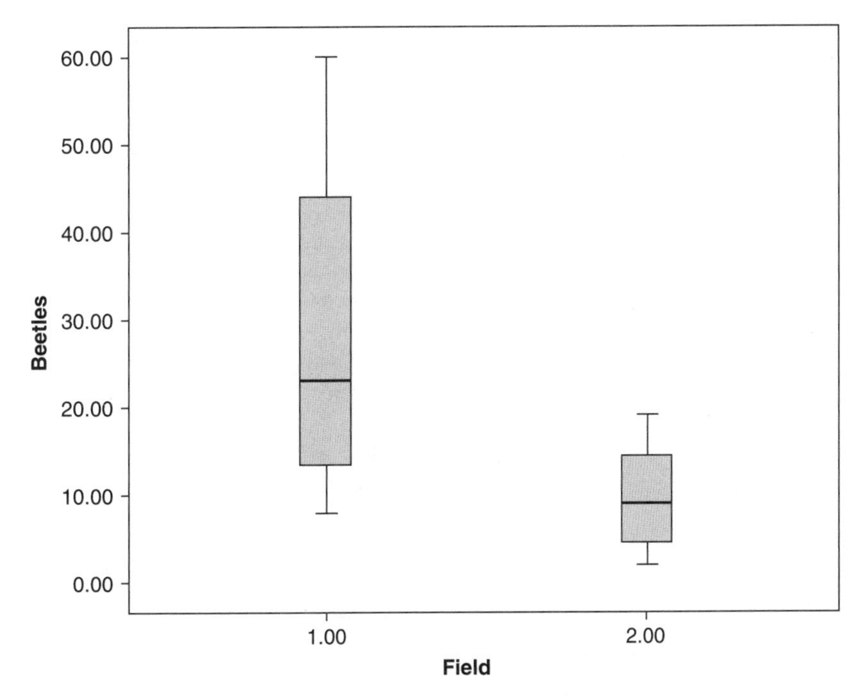

Figure 4.6 Box and whisker plot showing the numbers of beetles caught in traps in the two fields.

4.13 Self-assessment problems

Problem 4.1

The scores (in percent) of 25 students in a statistics test were as follows:

58	65	62	73	70	42	56	53	59	56	60	64	63
78	90	31	65	58	59	21	49	51	58	62	56	

Calculate the mean, standard deviation and standard error of the mean for these scores. The mean mark of students in finals exams is supposed to be 58%. Perform a one-sample *t* test to determine whether these students did significantly differ from expected.

Problem 4.2

The masses (in grams) of 16 randomly chosen tomatoes grown in a commercial glasshouse were as follows.

32	56	43	48	39	61	29	45
53	38	42	47	52	44	36	41

Other growers have found that the mean mass of this sort of tomato is 50 g. Perform a one-sample *t* test to determine whether the mean mass of tomatoes from this glasshouse is different from the expected mass. Give the 95% confidence intervals for the mean mass.

Problem 4.3

Students were tested on their ability to predict how moving bodies behave, both before and after attending a course on Newtonian physics. Their marks are tabulated here. Did attending the course have a significant effect on their test scores, and if so by how much?

	Before	After
Martha	45	42
Denise	56	50
Betty	32	19
Amanda	76	78
Eunice	65	63
Ivy	52	43
Pamela	60	62
Ethel	87	90
Letitia	49	38
Patricia	59	53

Problem 4.4

The pH of cactus cells was measured at dawn and at dusk using microprobes. The following results were obtained.

Dawn	5.3	5.6	5.2	7.1	4.2	4.9	5.4	5.7	6.3	5.5	5.7	5.6
Dusk	6.7	6.4	7.3	6.2	5.2	5.9	6.2	6.5	7.6	6.4	6.5	

(a) Using a statistical package, carry out a two-sample *t* test to determine if there is any significant difference in pH between the cells at these times.
(b) The cactus was identifiable and two sets of measurements were carried out on it. So why can't you analyse this experiment using the paired *t* test?

Problem 4.5

An experiment was carried out to investigate the effect of mechanical support on the yield of wheat plants. The masses of seed (in grams) produced by 20 control plants and 20 plants whose stems had been supported throughout their life were as follows:

Control	9.6	10.8	7.6	12.0	14.1	9.5	10.1	11.4	9.1	8.8
	9.2	10.3	10.8	8.3	12.6	11.1	10.4	9.4	11.9	8.6
Supported	10.3	13.2	9.9	10.3	8.1	12.1	7.9	12.4	10.8	9.7
	9.1	8.8	10.7	8.5	7.2	9.7	10.1	11.6	9.9	11.0

Using a statistical package, carry out a two-sample *t* test to determine whether the support has a significant effect on yield.

Problem 4.6

A population of 30 adult deer, which exhibit marked sexual dimorphism, were weighed at the start and at the end of summer, to investigate if they 'fatten up' to last over the subsequent winter. The following results (in kg) were obtained.

Deer	F1	F2	F3	F4	F5	F6	F7	F8	F9	F10	F11	F12	F13	F14	F15
Start	45	56	35	47	46	49	61	50	42	50	45	38	30	46	53
End	53	65	37	58	43	54	75	54	37	56	60	39	37	54	48

Deer	F16	F17	F18	M1	M2	M3	M4	M5	M6	M7	M8	M9	M10	M11	M12
Start	48	43	54	67	78	63	85	79	60	74	78	57	76	91	77
End	43	47	50	76	85	62	98	81	66	85	83	53	89	94	90

(a) Why is it not possible to transform this data to make it normally distributed?
(b) Carry out a Wilcoxon matched pairs test to determine if the animals had significantly different weight at the end compared with the start of summer.

Problem 4.7

In a behavioural experiment, scientists compared the amount of time that a macaque spent pacing back and forth (a sign of distress) when in a traditional cage compared with when it was in an 'environmentally enriched' cage. The animal was observed over 4 days for periods of 15 minutes every 2 hours, being moved at the end of each day to the other cage. The following results (in minutes spent pacing) were obtained.

Traditional cage:	0,	3,	1,	15,	0,	15,	0,	12,	1,	10,	0,	15
Enriched cage:	1,	0,	11,	0,	1,	0,	15,	1,	0,	0,	2,	1

Carry out a Mann–Whitney *U* test to see if the animal behaved differently in the two cages.

Problem 4.8

A new drug to improve wound healing was tested on students in Manchester. Tiny experimental lesions were made on their arms. Half were given the drug, while the other half received a placebo. Six weeks later the extent of scarring was assessed on an arbitrary scale ranging from 0 (no scar tissue) to 5 (heavy scarring). The following results were obtained.

Placebo:	1,	3,	2,	4,	3,	3,	2,	3,	3,	2,	3,	1,	0,	4,	3,	2,	3,	4,	3,	2
Drug:	1,	2,	2,	3,	0,	2,	0,	1,	2,	1,	2,	1,	4,	3,	0,	2,	2,	1,	0,	1

(a) Which test should you use to determine whether the drug had any effect on scarring?
(b) Carry out the test.

5 Testing for differences between more than two groups

ANOVA and its non-parametric equivalents

5.1 Introduction

We saw in the last chapter how you can use t tests and their non-parametric equivalents to compare one set of measurements with an expected value, or two sets of measurements with each other. However, there are many occasions in biology when we might want to save time and effort by comparing three or more groups.

- We might want to *compare* two or more groups of experimentally treated organisms or cells with controls. Do two sets of rats with mutations on separate chromosomes have *different* life expectancies from control 'wild-type' rats?
- We might want to *compare* three or more groups of organisms or cells. Do people on four drug regimes have different blood pressures? Do five strains of bacteria have different growth rates?
- We might want to *compare* three or more sets of related measurements or measurements repeated three or more times made on a single population. Are levels of aluminium in the same set of fish *different* at three or more times? Is the number of four species of birds recorded every hour at a feeding station *different* from each other?
- We might want to *compare* organisms or cells that have been influenced by two separate types of treatment. Do wheat plants given different levels of both nitrogen and phosphorous have *different* yields?

This chapter describes how you can use a set of tests called *analysis of variance* (ANOVA) to help determine whether there are differences and if so between which of the groups. Non-parametric equivalents for some of the tests will also be described. First, however, we must see why it is that you cannot use t tests for comparing multiple groups.

5.1.1 Why t tests are unsuitable

If you want to compare the means of more than two groups, you might think that you could simply compare each group with all the others using two-sample t tests. However, there are two good reasons why you should not do this. First, there is the problem of convenience. As the number of groups you are comparing goes up, the number of tests you must carry out rises rapidly, from 3 tests when comparing 3 groups to 45 tests for 10 groups. It would take a lot of time to do all these tests and it would be nearly impossible to present the results of all them!

Number of groups	3	4	5	6	7	8	9	10
Number of *t* tests	3	6	10	15	21	28	36	45

However, there is a second, more important problem. We reject a null hypothesis with 95% confidence, not 100% confidence. This means that in 1 in 20 tests we will falsely assume there is a significant difference between groups when none really exists (a type I error). If we carry out a lot of tests, the chances of making such an error go up rapidly, so if we carry out 45 tests there is about a 90% chance we will find significant effects even if none exist.

For these reasons you must use a rather different set of statistical tests to determine whether there is a difference between many groups; for normally distributed data you at least, you should use analysis of variance (**ANOVA**). ANOVA can be used to answer all the questions poised at the start of the chapter. All the ANOVA tests use similar logic, and it is perhaps easiest to see how they work by starting with the simplest form of the test, a **one-way ANOVA.**

ANOVA
Abbreviation for analysis of variance: a widely used series of tests which can determine whether there are significant differences between groups.

5.2 One-way ANOVA

5.2.1 Purpose

To test whether the means of two or more sets of **unrelated** measurements are different from each other. For instance it is used to test whether one or more groups of experimentally treated organisms are different from controls, or two or more species are different from one another.

5.2.2 Validity

Like *t* tests, one-way ANOVA is only valid if the data within each group is normally distributed. However, ANOVA also assumes that the variances of the groups are equal. Fortunately, this is not a strict condition, and research has shown that the variances of the groups can differ by factors of over four without it affecting the results of the test (Field, 2000).

5.2.3 The rationale of one-way ANOVA

One-way ANOVA works in a very different manner from *t* tests. Rather than examine the difference between the means directly, ANOVA looks at the **variability** of the data. Let's examine a simple example in which the means of the

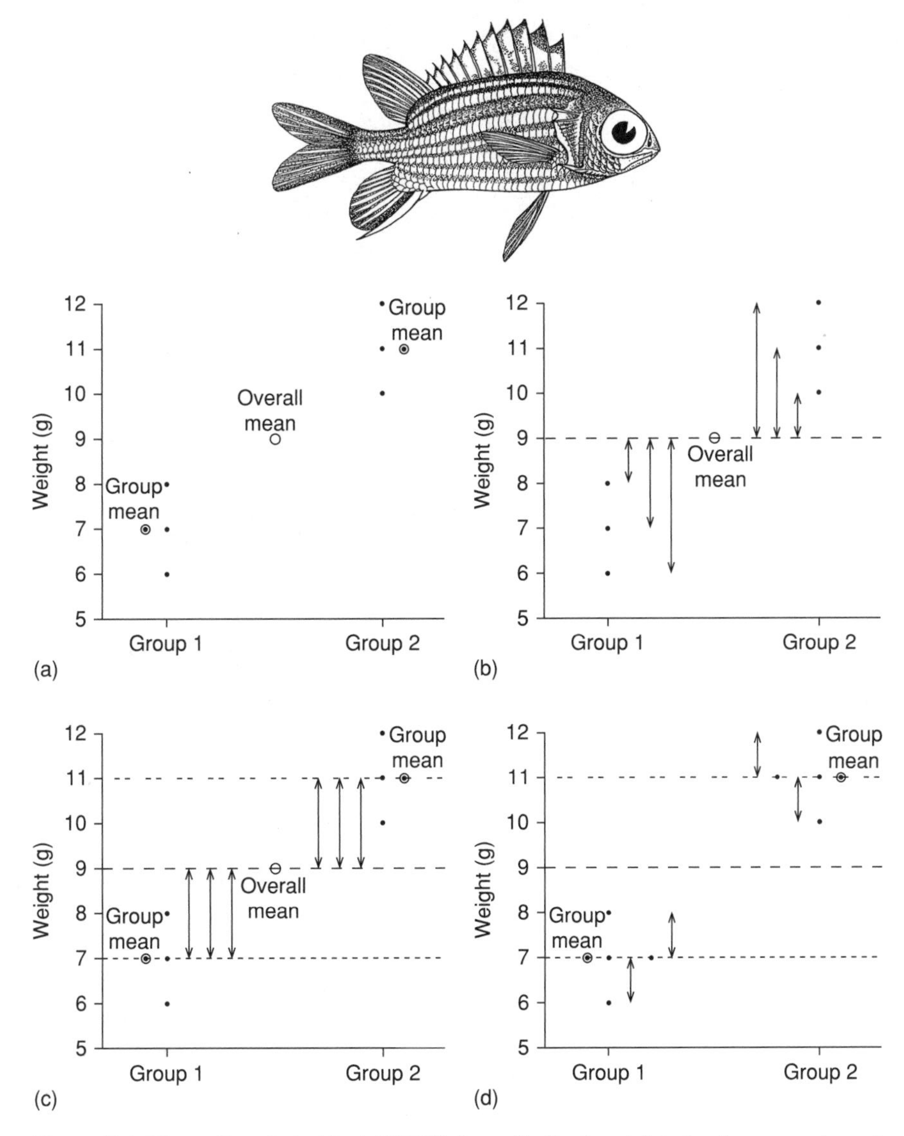

Figure 5.1 The rationale behind ANOVA: hypothetical weights for two samples of fish. **(a)** Calculate the overall mean and the group means. **(b)** The total variability is the sum of the squares of the distances of each point from the overall mean; this can be broken down into between-group variability and within-group variability. **(c)** The between-group variability is the sum of the squares of the distances from each point's group mean to the overall mean. **(d)** The within-group variability is the sum of the squares of the distances from each point to its group mean.

weights of just two small samples of fish are compared (Figure 5.1a). The overall variability is the sum of the squares of the distances from each point to the overall mean (Figure 5.1b); here it's $3^2 + 2^2 + 1^2 + 3^2 + 2^2 + 1^2 = 28$. But this can be split into two parts. First, there is the between-group variability, which is due to the differences between the group means. This is the sum of the squares of the distances of each point's group mean from the overall mean (Figure 5.1c);

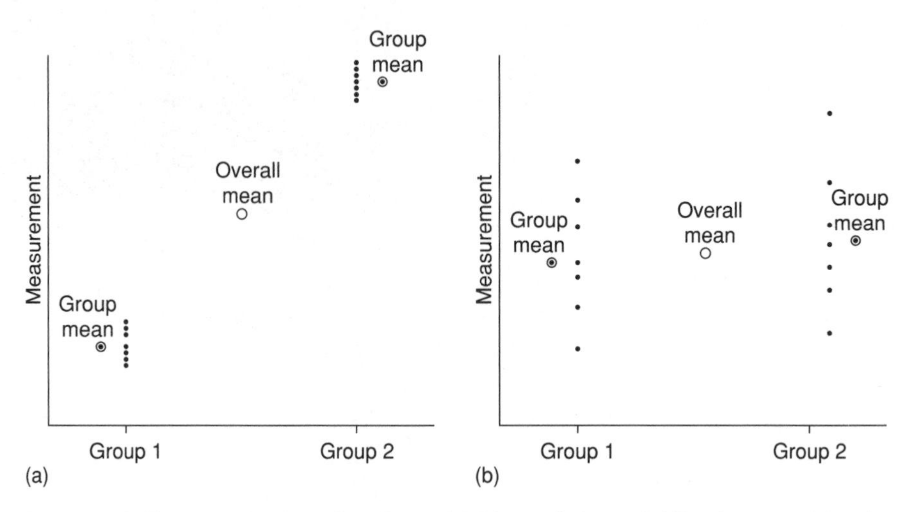

Figure 5.2 Two contrasting situations. (a) Most of the variability is caused by the group means being far apart. **(b)** Most of the variability is caused by differences within the groups.

here it's $(6 \times 2^2) = 24$. Second, there is the within-group variability, which is due to the scatter within each group. This is the sum of the squares of the distance from each point to its group mean (Figure 5.1d); here it's $(4 \times 1^2) + (2 \times 0^2) = 4$.

ANOVA compares the between-group variability and the within-group variability. To show how this helps, let's look at two contrasting situations. In Figure 5.2a the two means are far apart and there is little scatter within each group; the between-group variability will clearly be much larger than the within-group variability. In Figure 5.2b the means are close together and there is much scatter within each group; the between-group variability will be lower than the within-group variability.

The test statistic in ANOVA tests is the *F* statistic, a measure of the ratio of between-group to within-group variability. Calculating *F* is quite a long-winded process, however, and involves producing a table like the one shown below and in Example 5.1.

ANOVA

Fishweig

	Sum of squares	df	Mean square	*F*	Sig.
Between Groups	24.000	1	24.000	24.000	0.008
Within Groups	4.000	4	1.000		
Total	28.000	5			

(a) The first stage is to calculate the variabilities due to each factor to produce the so-called sums of squares (SS).

(b) You cannot directly compare sums of squares, because they are the result of adding up different numbers of points. The next stage is therefore to calculate

variance
A measure of the variability of data: the square of their standard deviation.

mean square
The variance due to a particular factor in analysis of variance (ANOVA).

degrees of freedom (DF)
A concept used in parametric statistics, based on the amount of information you have when you examine samples. The number of degrees of freedom is generally the total number of observations you make minus the number of parameters you estimate from the samples.

the actual **variance** or **mean squares** (MS) due to each factor. This is calculated by dividing each sum of squares by the correct number of **degrees of freedom.**

(i) If there are n groups, the between-group degrees of freedom $DF_B = n - 1$.
(ii) If there are N items in total and r items in each group, there will be $r - 1$ degrees of freedom in each group, hence $n(r - 1)$ in total. The within-group degrees of freedom, $DF_W = N - n$.
(iii) If there are N items in total, the total number of degrees of freedom $DF_T = N - 1$.

(c) The last stage is to calculate the test statistic F. This is the ratio of the between-group mean square MS_B to the within-group mean square MS_W.

$$F = MS_B/MS_W \quad (5.1)$$

The larger the value of F, the more likely it is that the means are significantly different.

5.2.4 Problems with names

Unfortunately, because ANOVA was developed separately by different branches of science, there are problems with the nomenclature of ANOVA; there are two synonyms for *between* and *within*:

between = treatment = factor
within = error = residual

You must be able to recognise all of them as different names may be used in scientific papers or different computer packages. Then you can cope with any statistics book or any lecturer!

5.2.5 Carrying out a one-way ANOVA test

The actual workings of ANOVA tests are actually quite complex, so you are best advised to use a computer package such as SPSS or MINITAB to perform them, but they involve the same four basic steps as the t tests we have already carried out. Once again the test is best described by running through a simple example, in this case of the fish shown in Figure 5.1.

Example 5.1

Mass of group 1:	6,	7,	8.
Mass of group 2:	10,	11,	12

Is the mass of the two groups of fish significantly different from each other?

Solution

Step 1: Formulating the null hypothesis

The null hypothesis is that the groups have the same mean. In this case the hypothesis is that the two groups of fish have the same mean weight.

Step 2: Calculate the test statistic

Using SPSS

Just as for the two-sample t test, when performing one-way ANOVA *all the data points should be entered into a single column.* Call it something like **fishweight**. To distinguish between the separate groups, you must then create a second, subscript, column with different integer values for each group (here 1 and 2) and give it a name (here **sample**). To carry out the test click on the **Analyze** menu, then move to the **Compare Means** bar and click on **One-Way ANOVA**. SPSS will come up with the **One-Way ANOVA** dialogue box.

Put the variable you are testing (here **fishweight**) into the **Dependent List** box and the factor (here **sample**) into the **Factor** box. You should also click on the **Options** box and tick the **Descriptives box** to get SPSS to calculate the means, standard deviations, and standard errors for each group. The completed data sheet and dialogue boxes are shown below.

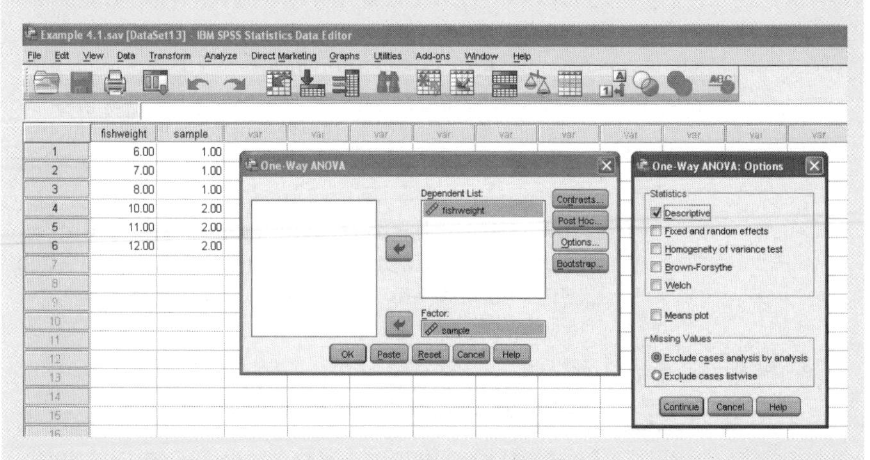

Click on **Continue** to get back to the original dialogue box and finally click on **OK** to start the test. SPSS will print the following results.

Oneway

Descriptives

fishweight

					95% Confidence interval for mean			
	N	Mean	Std. deviation	Std. error	Lower bound	Upper bound	Minimum	Maximum
1.00	3	7.0000	1.00000	0.57735	4.5159	9.4841	6.00	8.00
2.00	3	11.0000	1.00000	0.57735	8.5159	13.4841	10.00	12.00
Total	6	9.0000	2.36643	0.96609	6.5166	11.4834	6.00	12.00

ANOVA

Fishweig

	Sum of squares	df	Mean square	*F*	Sig.
Between Groups	24.000	1	24.000	24.000	0.008
Within Groups	4.000	4	1.000		
Total	28.000	5			

The first table gives the descriptive statistics for the two groups. The second table gives the completed ANOVA table which shows that $F = 24.00$.

Using MINITAB

As in SPSS I recommend putting all the data items into a single column and then creating a second subscript column with different integer values for each group (here 1 and 2). [An alternative is to put your groups into different columns and run the **One-Way (Unstacked)** test.] Next, click on the **Stat** menu, move onto the **ANOVA** bar, and click on **One-Way**. MINITAB will produce the **One-Way Analysis of Variance** dialogue box. Put the variable you want to test (here **fish weight**) into the **Response** box and the subscript column (here **sample**) into the **Factor** box to give the data set and completed box shown below.

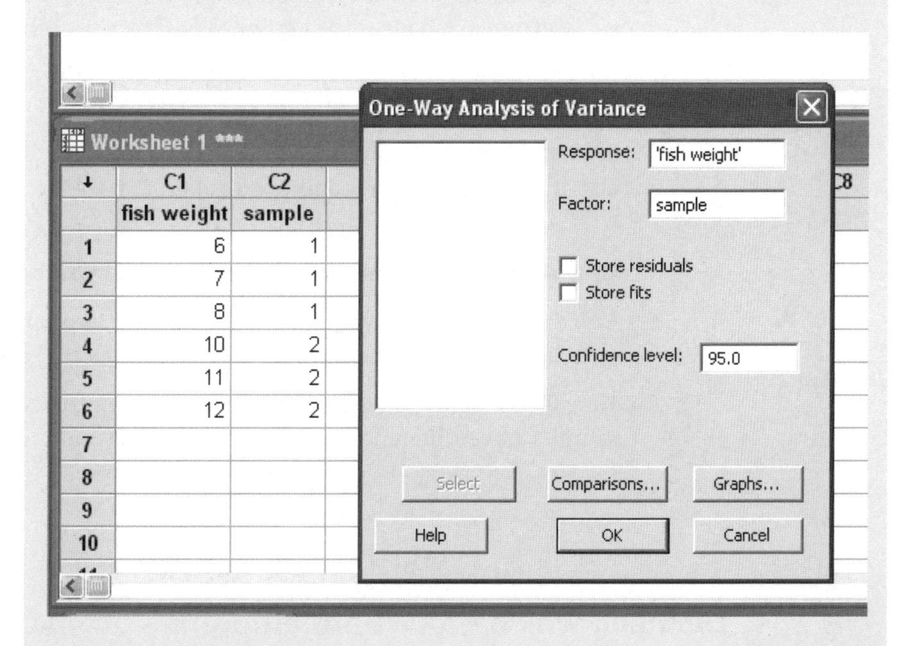

Click on **OK** to run the tests.

MINITAB comes up with the following results.

One-way ANOVA: fish weight versus sample

```
Source  DF     SS     MS      F      P

sample   1  24.00  24.00  24.00  0.008

Error    4   4.00   1.00

Total    5  28.00

S = 1   R-Sq = 85.71%   R-Sq(adj) = 82.14%

                              Individual 95% CIs For Mean Based on

                              Pooled StDev

Level  N    Mean   StDev  ---+---------+---------+---------+------

1      3   7.000   1.000  (-------*-------)

2      3  11.000   1.000                    (-------*-------)

                          ---+---------+---------+---------+------

                             6.0       8.0      10.0      12.0

Pooled StDev = 0.4925
```

The first table gives the completed ANOVA table which shows that $F = 24.00$. The second gives the descriptive statistics for the two groups.

Step 3: Calculating the significance probability

SPSS and MINITAB automatically calculate not only the test statistic F, but also the significance probabilities Sig. and P. Here Sig. $= P = 0.008$.

Step 4: Deciding whether to reject the null hypothesis

- If Sig. or $P \leq 0.05$ you must reject the null hypothesis.
- If Sig. or $P > 0.05$ you have no evidence to reject the null hypothesis.

Here $0.008 < 0.05$ so we can reject the null hypothesis and say that the fish samples have different mean weights. In fact sample two is significantly heavier.

5.3 Deciding which groups are different – post hoc tests

The problem with the basic ANOVA test is that though it tells us whether there are differences between groups, it doesn't tell us which groups are different. This does not matter with the fish samples because there were only two groups,

post hoc tests
Statistical tests carried out if an analysis of variance is significant; they are used to determine which groups are different from each other.

but it will be a problem if you have three or more groups. Fortunately, statisticians have worked out several different **post hoc tests** that you can use to see which groups are different from each other, *but only if the ANOVA is itself significant.* They basically all make allowances for the problems caused by the fact that you are running several comparisons, but they do it in different ways.

SPSS and MINITAB both allow you to perform several of these post hoc tests, and different ones can be used depending on what you want to test.

- If you want to compare each group with all the others, the tests most used by biologists are the **Tukey** test and the **Scheffe** test.
- If you want to compare experimental groups with a control, then the test to use is the **Dunnett** test.

Let's have a look at a typical example, to see how to perform one way ANOVA and a relevant post hoc test.

Example 5.2

The effect of three different antibiotics on the growth of a bacterium was examined by adding them to Petri dishes, which were then inoculated with the bacteria. The diameter of the colonies (in millimetres) was then measured after three days. A control where no antibiotics were added was also included. The following results were obtained.

Control	4.7	5.3	5.9	4.6	4.9	5.0	5.3	4.2
	5.7	5.3	4.6	5.8	4.7	4.9		
Antibiotic A	3.5	4.6	4.4	3.9	3.8	3.6	4.1	4.3
	4.3	4.8	4.1	5.0	3.4	4.3		
Antibiotic B	4.7	5.2	5.4	4.4	6.1	4.8	5.3	5.5
	4.7	5.2						
Antibiotic C	4.3	5.7	5.3	5.6	4.5	4.9	5.1	5.3
	4.7	6.3	4.8	4.9	5.2	5.4	4.8	5.0

Carry out a one-way ANOVA test to determine whether any of the antibiotic treatments affected the growth of the bacteria and if so, which ones.

Solution

Step 1: Formulating the null hypothesis

The null hypothesis is that there was no difference in the mean diameters of the four groups of bacteria.

Step 2: Calculating the test statistic

In SPSS or MINITAB carry out the basic one-way ANOVA test as shown above.

SPSS will print the following results:

Descriptives

diameter

					95% Confidence interval for mean			
	N	Mean	Std. deviation	Std. error	Lower bound	Upper bound	Minimum	Maximum
1.00	14	5.0643	0.50476	0.13490	4.7728	5.3557	4.20	5.90
2.00	14	4.1500	0.47677	0.12742	3.8747	4.4253	3.40	5.00
3.00	10	5.1300	0.49453	0.15638	4.7762	5.4838	4.40	6.10
4.00	16	5.1125	0.49379	0.12345	4.8494	5.3756	4.30	6.30
Total	54	4.8537	0.63713	0.08670	4.6798	5.0276	3.40	6.30

ANOVA

diameter

	Sum of squares	df	Mean square	*F*	Sig.
Between groups	9.389	3	3.130	12.905	0.000
Within groups	12.126	50	0.243		
Total	21.514	53			

While MINITAB will give

One-way ANOVA: diameter versus treatment

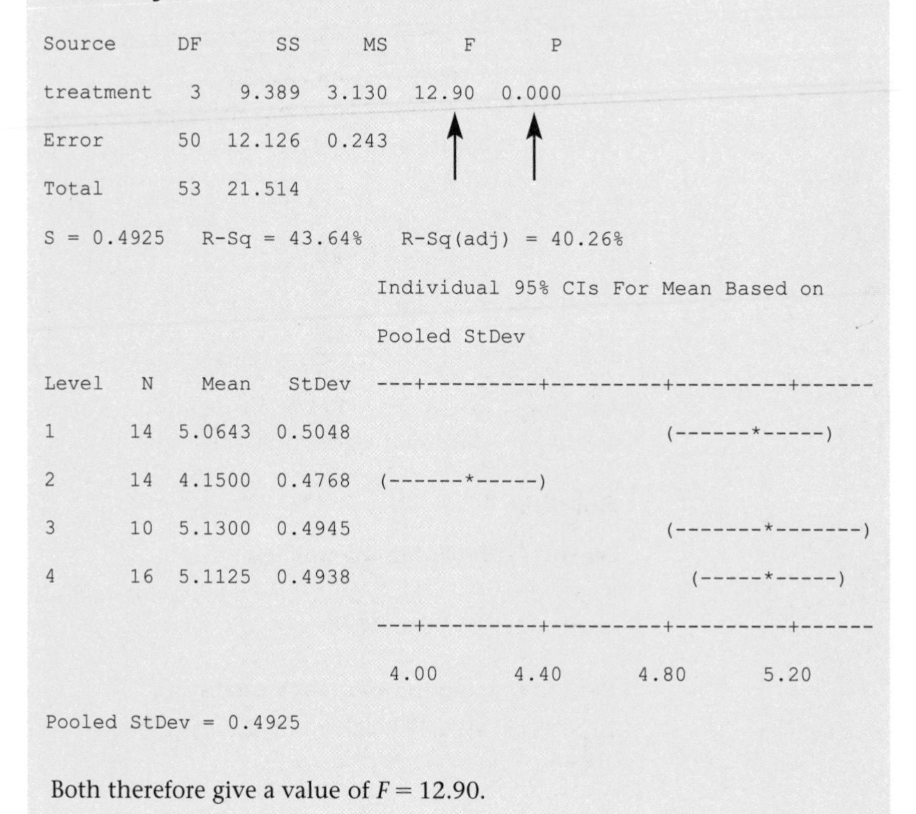

```
Source      DF      SS     MS      F      P
treatment    3   9.389  3.130  12.90  0.000
Error       50  12.126  0.243
Total       53  21.514
S = 0.4925   R-Sq = 43.64%   R-Sq(adj) = 40.26%

                              Individual 95% CIs For Mean Based on
                              Pooled StDev
Level   N    Mean   StDev   ---+---------+---------+---------+------
1      14  5.0643  0.5048                        (------*-----)
2      14  4.1500  0.4768   (------*-----)
3      10  5.1300  0.4945                        (-------*-------)
4      16  5.1125  0.4938                          (-----*-----)
                            ---+---------+---------+---------+------
                             4.00      4.40      4.80      5.20
Pooled StDev = 0.4925
```

Both therefore give a value of $F = 12.90$.

Step 3: Calculating the significance probability

SPSS and MINITAB both calculate that the significance probability Sig. = $P = 0.000$.

Step 4: Deciding whether to reject the null hypothesis

Here $0.000 < 0.05$ so we can reject the null hypothesis and say that there are significant differences between the mean diameters of the bacteria.

Step 5: Deciding which groups are different

The aim of the experiment was to test which if any of the treatments altered the diameter of the bacterial colonies, so we need to compare *each of treatments against the control*. The test to use is the Dunnett test.

Using SPSS

Repeat the ANOVA test but click on **Post Hoc** to reveal the **Post Hoc Multiple Comparisons** dialogue box. Tick the **Dunnett** box. The control is given by the subscript 1, so you need to change the **Control Category** from **Last** to **First**. The completed dialogue box is shown below.

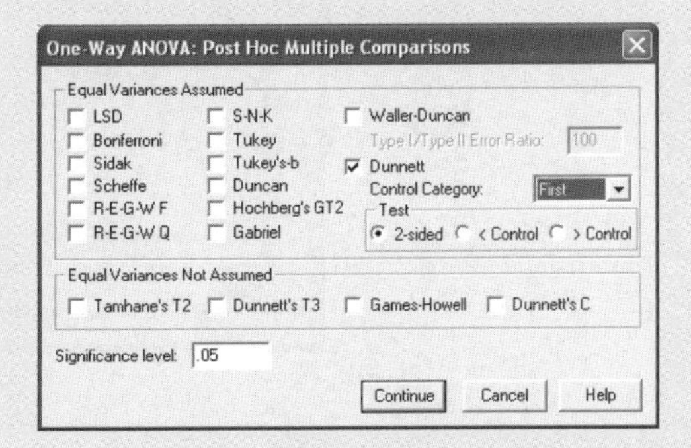

Finally, click on **Continue** and **OK** in the main dialogue box. As well as the results for the ANOVA SPSS will produce the following table.

Post Hoc tests

Multiple comparisons

Dependent variable: diameter
Dunnett *t* (two-sided)[a]

(*I*) treatment	(*J*) treatment	Mean difference (*I*–*J*)	Std. error	Sig.	95% Confidence interval	
					Lower bound	Upper bound
2.00	1.00	−0.91429*	0.18613	0.000	−1.3658	−0.4628
3.00	1.00	0.06571	0.20390	0.978	−0.4289	0.5603
4.00	1.00	0.04821	0.18022	0.987	−0.3890	0.4854

*The mean difference is significant at the 0.05 level.
[a] Dunnett *t*-tests treat one group as a control, and compare all other groups against it.

The important column is the **Sig.** Column. This tells you that of the three treatment groups, 2, 3 and 4, only 2 has a significantly different mean (Sig. = 0.000 which is less than 0.05) from the control. At 4.15 mm its mean diameter is almost 1 mm less than that of the control (5.06 mm). In groups 3 and 4 the significance probabilities (0.978 and 0.987) are well over 0.05, so they are not different from the control.

Using MINITAB

Repeat the ANOVA test but click on **Comparisons** to reveal the **One-Way Multiple Comparisons** dialogue box. Tick the **Dunnett** box. The control is given by the subscript 1, so you need to enter 1 into the **Control Group Level** box. The completed dialogue box is shown below.

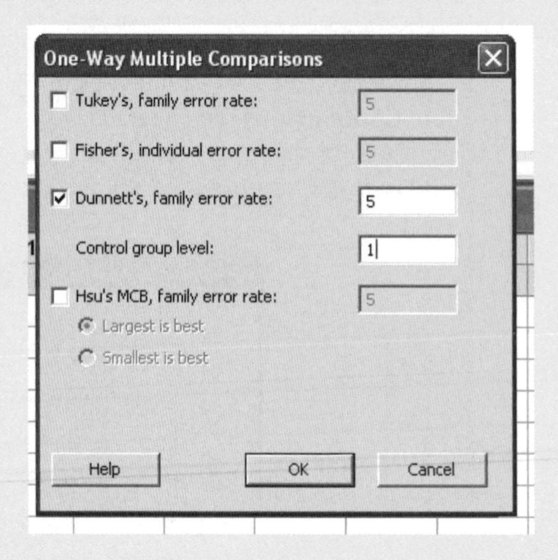

Finally, click on **OK**, then **OK** in the main dialogue box. As well as the results for the ANOVA MINITAB will produce (amongst other information) the following table.

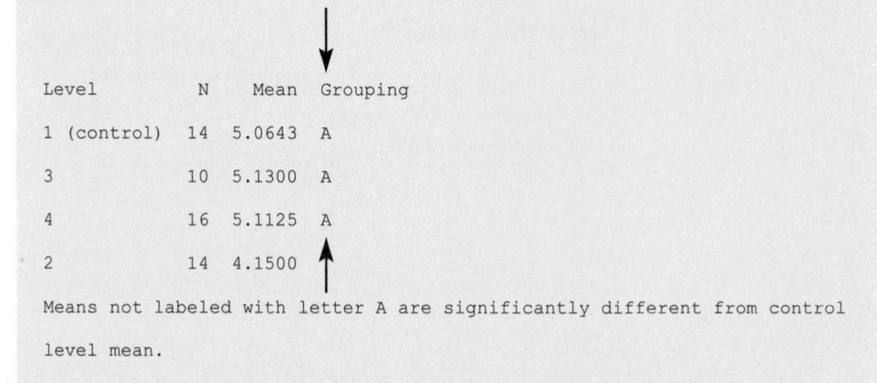

Level	N	Mean	Grouping
1 (control)	14	5.0643	A
3	10	5.1300	A
4	16	5.1125	A
2	14	4.1500	

Means not labeled with letter A are significantly different from control level mean.

This tells you that of the three groups only group 2 has a mean significantly different from that of the control group 1.

5.4 Presenting the results of one-way ANOVAs

If you are comparing more than two groups, it is best to present your results in the form of a bar chart. For instance the results of Example 5.2 can be presented as in Figure 5.3a or b.

Asterisks should be used, as shown in Figure 5.3a, if you want to emphasise whether individual groups are different from a control, for instance if you have carried out a Dunnett post hoc test on your data. Here only antibiotic A is

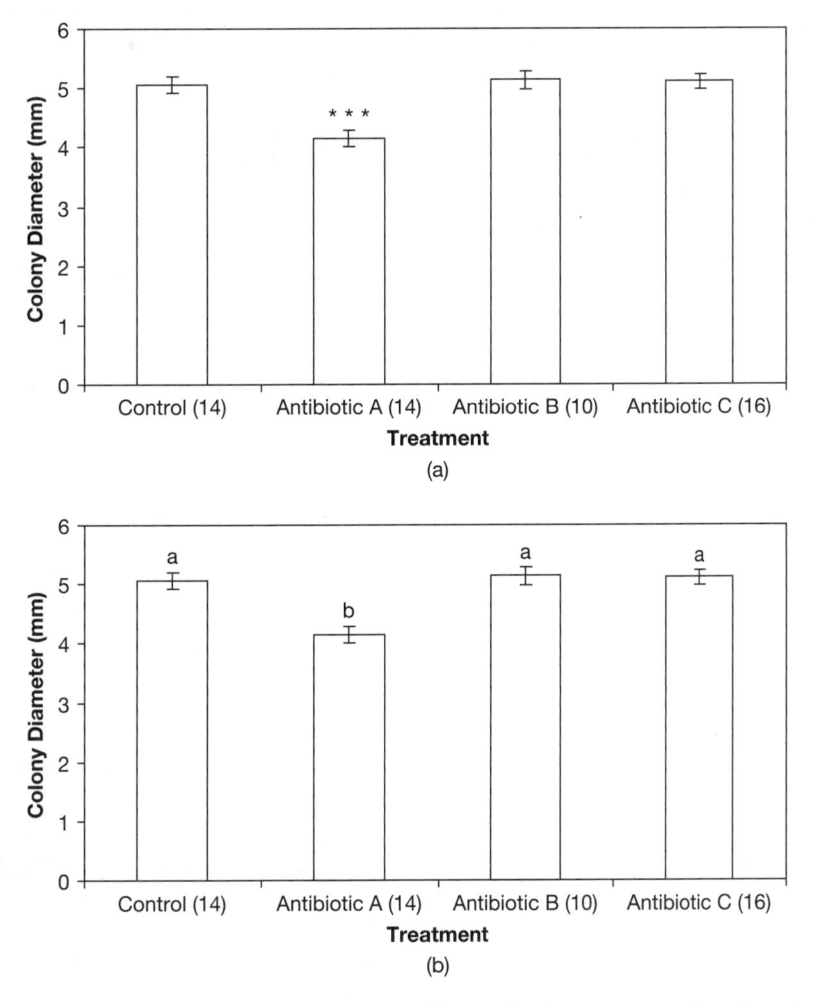

Figure 5.3 Bar chart showing the means with standard error bars of the diameters of bacterial colonies subjected to different antibiotic treatments. (a) Numbers in brackets give sample size. Asterisks denote the degree of significance of differences of diameter compared with controls: * $P < 0.05$; ** $P < 0.01$; *** $P < 0.001$. (b) Numbers in brackets give sample size. Letters denote significant differences between groups. Groups denoted by the same letter are not significantly different from each other.

significantly different from the control. The letter notation (Figure 5.3b) is preferable if you want to show which groups are different from each other and if you have performed a Tukey or Scheffe test to do just that. In this example colonies given antibiotic A were significantly different from all the others.

Once again, you should refer to the figure in the text of your results section, saying something like

> The mean colony diameters under the different antibiotic treatments are shown in Figure 5.3a. A one-way ANOVA showed that there were significant differences between antibiotic treatments ($F_3 = 12.90$), while a Dunnett's test showed that colonies given antibiotic A were significantly smaller than the controls.

5.5 Repeated measures ANOVA

5.5.1 Purpose

To test whether the means of a two or more sets of **related** measurements are different from each other. For instance it is used to test whether one group of experimentally treated organisms are different at several times before and after a treatment or if two or more sets of measurements taken at known time points are different from each other.

5.5.2 Rationale

Repeated measures ANOVA acts in the same way as one-way ANOVA, but it improves the chances of detecting differences between groups by removing the within group variability, just as a paired *t* test does.

5.5.3 Validity

Like *t* tests, repeated measures ANOVA is only valid if the data within each group is normally distributed. It also assumes that the variances of the groups are equal, though this condition is not strict.

5.5.4 Carrying out repeated measures ANOVA test

5.5.4.1 Using SPSS

Carrying out repeated measures ANOVA tests in SPSS is even more long-winded process than doing one-way ANOVA, so it is best demonstrated by an example.

Example 5.3

In an experiment to investigate the time course of the effect of exercise on the rate of sweating in soldiers in the desert, the following results were obtained.

Soldier	1	2	3	4	5	6	7	8	9	10
Rate before (litres/hour)	3.6	3.9	4.2	4.0	3.8	3.5	4.2	4.0	3.9	3.8
During	4.5	4.4	4.8	4.3	4.6	4.5	5.0	4.6	4.1	4.6
1 hour after	3.9	4.4	3.7	3.9	3.5	4.2	4.0	4.1	3.6	4.6

Carry out a repeated measures ANOVA to find out whether the rate of sweating altered during exercise and afterwards compared with before.

Step 1: Formulating the null hypothesis

The null hypothesis is that there was no difference in the rate of sweating between the times the measurements were taken.

Step 2: Calculating the test statistic

In SPSS enter the measurements of sweating rate into **three separate columns,** making sure the results from each soldier is put into the same row each time. Call the columns, say **before, during** and **after.**

To carry out the test click on the **Analyze** menu, then move to the **General Linear Model** bar and click on **Repeated Measures.** SPSS will come up with the **Repeated Measures Define Factor(s)** dialogue box. You first need to tell the computer the name of the factor that might affect the results. Here it's time before during and after exercise, so type **time** into the **Within Subject Factor Name** box. Next, you need to tell the computer how many experimental conditions or *levels* there were. Here there were 3 (before, during and after), so type 3 into the **Number of Levels** box. Next, click on the **Add** tab to input this data. The computer will print time(3) in the large box. The completed table is shown below.

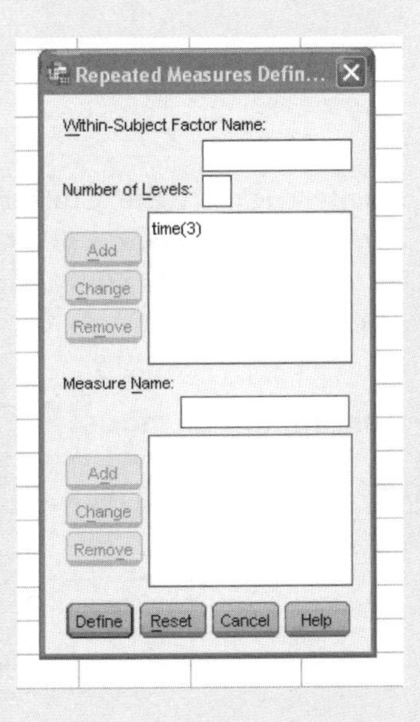

Now click on the **Define** tab to get into the main **Repeated Measures** dialogue box. Next, you must tell the computer which are the three Within Subject Variables (time). To do this, click on each of the three columns — before, during and after — in turn and clicking on the top arrow. The data is now entered.

To get other useful things done you should also click on the **Options** tab. This brings up the **Repeated Measures: Options** dialogue box. Click onto **descriptives,** to get the means and standard deviations. Unfortunately there is no Dunnett test to compare groups with a control, but you can carry out tests that compare each group with all the others. Other authors who know much more about this (see Field, 2000) than myself suggest that the **Bonferroni** is the most reliable post hoc test to use for repeated measures ANOVA. To perform it click on **time** within the **estimated marginal means** box and move it with the arrow into the **Display means for:** box. Now you can tick the **Compare Main Effects** and change the **confidence interval adjustment** box from LSD (none) to **Bonferroni.** The data and completed boxes are shown below.

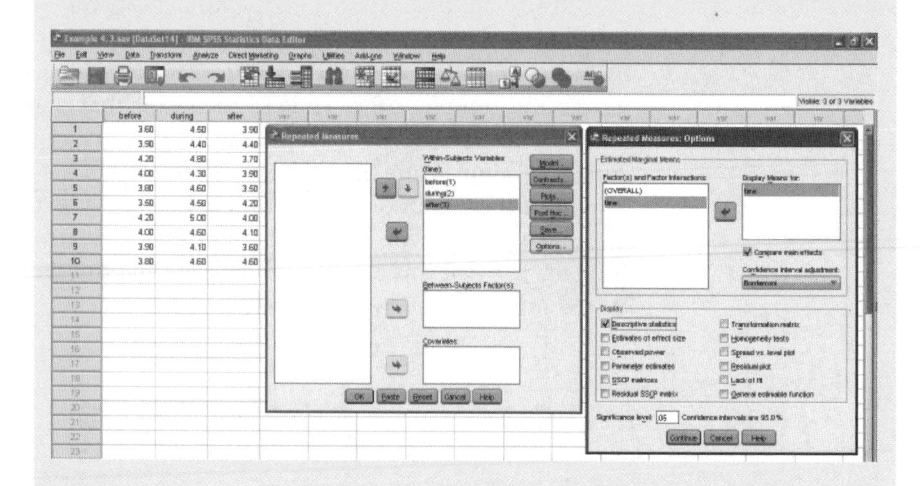

Finally click on the **Continue** tab, and when the main dialogue box appears again click on **OK** to run the test. SPSS comes up a huge mass of results. However, only the ones that are important for us are shown below.

General linear model

Descriptive statistics

	Mean	Std. deviation	*N*
before	3.8900	0.22828	10
during	4.5400	0.25033	10
after	3.9900	0.34785	10

Mauchly's test of sphericity[b]

Measure: MEASURE_1

Within subject effect	Mauchly's W	Approx. Chi-Square	df	Sig.	Epsilon[a] Greenhouse-Geisser	Huynh-Feldt	Lower-bound
Time	0.681	3.069	2	0.216	0.758	0.880	0.500

Tests the null hypthesis that the error covariance matrix of the orthonormalized transformed dependent variables is proportional to an identity matrix.

[a] May be used to adjust the degrees of freedom for the averaged tests of significance. Corrected tests are displayed in the Tests of Within-Subjects Effects table.

[b] Design: Intercept
Within Subjects Design: time

Tests of within-subjects effects

Measure: MEASURE_1

Source		Type III sum of squares	df	Mean square	F	Sig.
Time	Sphericity assumed	2.450	2	1.225	16.662	0.000
	Greenhouse–Geisser	2.450	1.517	1.615	16.662	0.000
	Huynh–Feldt	2.450	1.760	1.392	16.662	0.000
	Lower–bound	2.450	1.000	2.450	16.662	0.003
Error(time)	Sphericity assumed	1.323	18	0.074		
	Greenhouse–Geisser	1.323	13.650	0.097		
	Huynh–Feldt	1.323	15.836	0.084		
	Lower–bound	1.323	9.000	0.147		

Pairwise comparisons

Measure: MEASURE_1

(I) time	(J) time	Mean difference (I–J)	Std. error	Sig.[a]	95% Confidence interval for difference[a] Lower bound	Upper bound
1	2	−0.650*	0.082	0.000	−0.891	−0.409
	3	−0.100	0.144	1.000	−0.522	0.322
2	1	0.650*	0.082	0.000	0.409	0.891
	3	0.550*	0.129	0.006	0.171	0.929
3	1	0.100	0.144	1.000	−0.322	0.522
	2	−0.550*	0.129	0.006	−0.929	−0.171

Based on estimated marginal means
* The mean difference is significant at the 0.05 level.
[a] Adjustment for multiple comparisons: Bonferroni.

The first thing you must check is whether the data passes Mauchley's sphericity test in the second table. If the data shows significant non-sphericity Sig. < 0.05. Here, fortunately, Sig. $= 0.216$, so we can go along to examine the F ratio for **Sphericity Assumed** which is shown in the **Tests of Within-Subjects Effects** box. Here $F = 16.662$.

Using MINITAB

There is no way of performing a Repeated measures ANOVA in MINITAB.

Step 3: Calculating the significance probability

SPSS calculates that the significance probability Sig. = 0.000.

Step 4: Deciding whether to reject the null hypothesis

- If Sig ≤ 0.05 you must reject the null hypothesis.
- If Sig > 0.05 you have no evidence to reject the null hypothesis.

Here Sig. = 0.000 < 0.05 so we can reject the null hypothesis and say that there are significant differences between the mean rates of sweating at the three times.

Step 5: Deciding which groups are different

The aim of the experiment was to test when was the rate of sweating different from before exercise. Here during (group 2) vs before (group 1) has Sig. = 0.000, so clearly the soldiers sweated more during exercise. However, after (group 3) vs before (group 1) has Sig. = 1.000, so clearly soldiers didn't have significantly different rates of sweating after exercise from before.

NB I have just covered the very basics of this test. To find out about the theory behind repeated measures ANOVA, and the meaning and importance of sphericity, you are advised to go to more comprehensive books such as Field (2000).

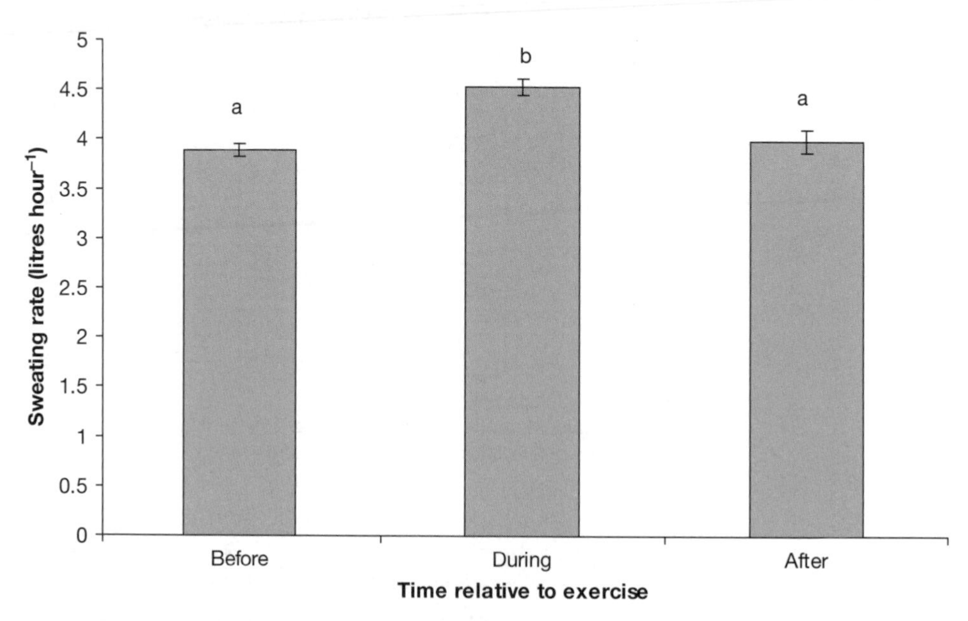

Figure 5.4 Mean sweating rates of soldiers before, during and after exercise. For all groups $n = 9$. Letters denote significant differences between groups. Groups denoted by the same letter are not significantly different from each other.

5.5.5 Presenting the results of repeated Measures ANOVA

Like one-way ANOVA, the best way to present the results of repeated measures ANOVA is in a bar chart showing means at each time, with standard error bars. Letters above the bars show which groups are different from the others. Note that because repeated measures removes the within-group variability, it is possible for means to be significantly different even if the error bars are overlapping. As an example the results for the sweating rates of soldiers are shown below.

Describe the results saying something like

> The results for the experiments on the sweating rates of soldiers are shown in Figure 5.4. A repeated measures ANOVA showed that sweating rates were significantly different at the three times ($F_2 = 16.62$). Bonferroni post hoc tests showed that rates were significantly higher only during exercise.

5.6 The Kruskall–Wallis test

5.6.1 Purpose

To test whether the medians of two or more sets of **unrelated** measurements are different from each other. It is therefore the non-parametric version of one-way ANOVA. For instance it is used to test whether one or more groups of experimentally treated organisms are different from controls, or two or more species are different from one another.

5.6.2 Rationale

The Kruskall–Wallis test starts in the same way as the Mann–Whitney *U* test, by assigning each observation its rank within all the measurements. If there are tied ranks, each is assigned the average value. The sum of the ranks in each sample, *R*, is then calculated. Finally the test statistic, *K*, is calculated using the formula

$$K = [\Sigma(R^2/n) \times 12/N(N + 1)] - 3(N + 1) \qquad (5.2)$$

Where n is the size of each sample and N is the total number of observations. The more different the medians of the samples are, the larger will be the sum of R^2/n, so the bigger *K* will be. In this test, the null hypothesis is rejected if *K* is **greater than or equal to** a critical value.

5.6.3 Carrying out the test

The method of carrying out the test is best seen by working through an example.

Example 5.4

In a student project to investigate the effectiveness of different teaching methods, three groups of schoolchildren were tested on the subject after working through different Computer-aided learning (CAL) packages. The results were as follows.

Package 1: 12, 16, 4, 10, 8, 15, 19, 3, 18, 5.

Package 2: 9, 15, 18, 2, 16, 7, 3, 17, 12, 15.

Package 3: 10, 15, 5, 3, 16, 9, 15, 13, 20, 10.

Using the **Explore** command, SPSS tells us that the median scores were 11, 10.5 and 11.5, but are these significantly different?

Solution

Step 1: Formulating the null hypothesis

The null hypothesis is that there is no difference in the median scores of the three groups.

Step 2: Calculating the test statistic

Using a calculator

You must first work out the ranks of all the observations. These are given in brackets in the table below.

Package 1	Package 2	Package 3
12 (15.5)	9 (10.5)	10 (13)
16 (24)	15 (20)	15 (20)
4 (5)	18 (27.5)	5 (6.5)
10 (13)	2 (1)	3 (3)
8 (9)	16 (24)	16 (24)
15 (20)	7 (8)	9 (10.5)
19 (29)	3 (3)	15 (20)
3 (3)	17 (26)	13 (17)
18 (27.5)	12 (15.5)	20 (30)
5 (6.5)	15 (20)	10 (13)

n	10	10	10
R	156.5	146	162.5
R^2	24492.25	21316	26406.25
R^2/n	2449.2	2131.6	2640.6

$K = [(2449.2 + 2131.6 + 2640.6) \times 12/(30 \times 31)] - (3 \times 31)$

$K = 93.18 - 93 = 0.18$

Using SPSS

As for one-way ANOVA, you must first put all the data (called, say, **testscore**) into the same column because each measurement is on a different schoolchild. To distinguish between the three packages, you must create a second, subscript, column (called, say, **CALtype**) with one of three values, here 1, 2 and 3. Simply click on the **Analyze** menu, move onto the **Nonparametric tests** bar, onto **Legacy Dialogs** and click onto the **k Independent Samples** bar. SPSS will come up with the **Tests for Several Independent Samples** dialogue box. Put the variable you want to test into the **Test Variable** box, making sure the **Kruskall–Wallis** test type is ticked. Put the subscript column into the

Grouping Variable box. Define the range by clicking on the **Define Range** tab to reveal the dialogue box and putting in the minimum and maximum values of the subscript column (here 1 and 3) in the **Range for grouping variable** boxes. The completed dialogue boxes and data are shown below.

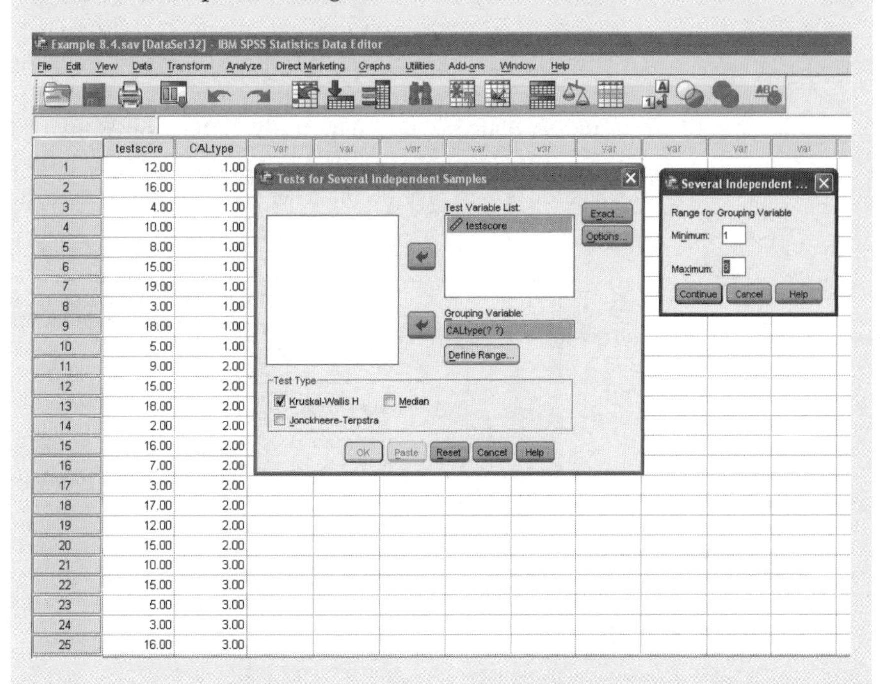

Finally, click on **Continue** and **OK** to run the test. SPSS will come up with the following results.

Ranks

	CALtype	*N*	Mean rank
testscore	1.00	10	15.65
	2.00	10	14.60
	3.00	10	16.25
	Total	30	

Test statistics[a,b]

	Test Score
Chi-square	0.181 ←
df	2
Asymp. sig.	0.913 ←

[a] Kruskal–Wallis test.

[b] Grouping variable: CALtype

SPSS calls K chi-square and gives the value as 0.181.

NB SPSS now has new analysis available for this test, available through the **Independent Samples** bar. I don't recommend this new method, however, as although it carries out the test it doesn't actually present you with any statistics!

Using MINITAB

As for one-way ANOVA, you must first put all the data (called, say, **test score**) into the same column because each measurement is on a different schoolchild. To distinguish between the three packages, you must create a second, subscript, column (called, say, **CAL type**) with one of three values, here 1, 2 and 3. Next, click on the **Stat** menu, move onto the **Nonparametrics** bar, and click on **Kruskall–Wallis**. MINITAB will produce the **Kruskall–Wallis** dialogue box. Put the variable you want to test (here **test score**) into the **Response** box and the subscript column (here **CAL type**) into the **Factor** box to give the data set and completed box shown below.

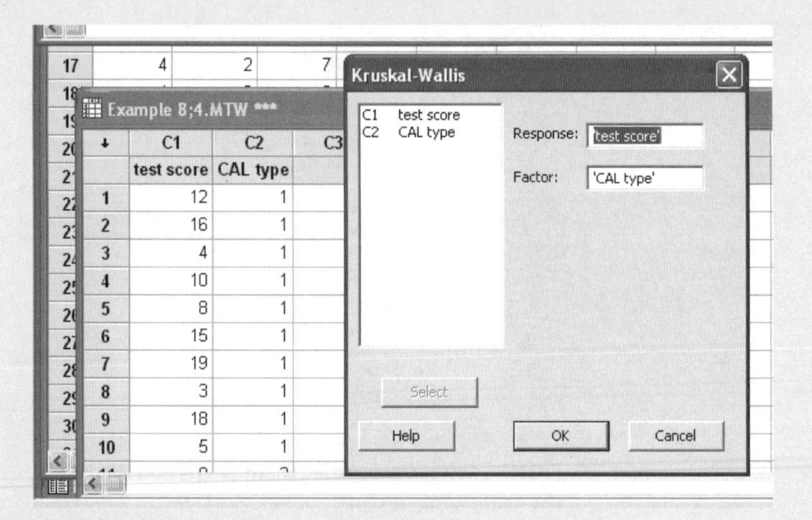

Click on **OK** to run the test.

MINITAB comes up with the following results.

Kruskal–Wallis Test: test score versus CAL type

```
Kruskal-Wallis Test on test score

CAL type    N  Median  Ave Rank      Z

1          10   11.00      15.3  -0.11

2          10   13.50      15.6   0.02

3          10   11.50      15.7   0.09

Overall    30              15.5

H = 0.01  DF = 2  P = 0.993
H = 0.01  DF = 2  P = 0.993  (adjusted for ties)
```

MINITAB doesn't actually give a value for *K*, but never mind.

Step 3: Calculating the significance probability

You must calculate the probability P that the test statistic K would be **greater than or equal to** K if the null hypothesis were true.

Using a calculator

You must compare your value of K with the critical value of the χ^2 statistic for $(G - 1)$ degrees of freedom, where G is the number of groups. This is given in Table S3 at the end of the book.

Looking up in the chi-square distribution for (3–1) = 2 degrees of freedom gives a critical value of chi-square = 5.99.

Using SPSS and MINITAB

SPSS shows that the probability Asymp. Sig. (two-tailed) of getting such test statistics is 0.993. MINITAB also gives a value for P of 0.993 (use the value in the lower line which adjusts for ties).

Step 4: Deciding whether to reject the null hypothesis

Using a calculator

- If χ^2 is greater than or equal to the critical value, you must reject the null hypothesis. You can say that the medians of the samples are significantly different from each other.
- If χ^2 is less than the critical value, you cannot reject the null hypothesis. You can say that there is no significant difference between the medians of the samples.

Here $\chi^2 = 0.181 < 5.99$.

Using SPSS and MINITAB

- If Asymp. Sig. (two-tailed) pr $P \leq 0.05$ you must reject the null hypothesis. Therefore you can say that the medians of the samples are significantly different from each other.
- If Asymp. Sig. (two-tailed) or $P > 0.05$, you have no evidence to reject the null hypothesis. Therefore you can say that there is no significant difference between the medians of the samples.

Here Asymp. Sig. (two-tailed) = $P = 0.993 > 0.05$.

Therefore we have no evidence to reject the null hypothesis. The children who took the different CAL packages did not have significantly different median scores.

5.6.4 Post hoc tests for Kruskall–Wallis

Of course, just as for one-way ANOVA, if there is a significant difference between the groups, you may want to know which groups are different from each other. To do this it is best to carry out separate Mann–Whitney U tests for each test

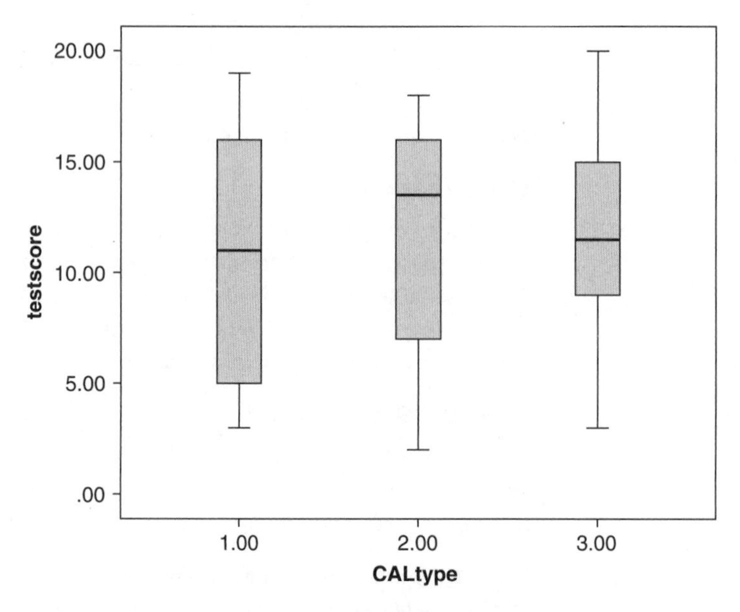

Figure 5.5 Box and whisker plot showing the medians, quartiles and range of the test scores of children who had taken different CAL packages. For all samples $n = 10$.

you want to compare, but use the Dunn–Sidak correction so that the significance probability for each test, rather than being 0.05 is given by the equation $P = (1 - 0.95^{1/k})$ where k is the number of tests.

5.6.5 Presenting the results of the Kruskall–Wallis test

If you are comparing more than two groups, it is best to present your results in the form of a box and whisker plot. For instance the children's test results can be presented as shown in Figure 5.5.

To show which groups (if any) are different from each other, it is best to use the letter notation as used for Tukey test (Figure 5.3b). Once again, you should refer to the figure in the text of your results section, saying something like

> The median mark of the children who had been given the different CAL packages are shown in Figure 5.5. A Kruskall–Wallis tests showed that there was no significant difference between the median scores.

5.7 The Friedman test

5.7.1 Purpose

To test whether the means of two or more sets of **related** measurements are different from each other. For instance it could be used to test whether one group of experimentally treated organisms are different at several times after a treatment, or if two or more sets of measurements taken at known time points are different from each other. Friedman's test is the non-parametric equivalent of repeated measures ANOVA.

5.7.2 Rationale

Take the case of an investigation looking at whether there are differences in a measurement between four time points, in 10 experimental animals. The measurements within each of the b blocks are first given ranks. In this case, as there are 10 animals measured there will be 10 blocks. The ranks are then summed for each of the a groups. As the animals are measured four times there will be four groups. The sums for each group are given the term R_i. The test statistic, χ^2, is then calculated using the following formula.

$$\chi^2 = \frac{12\Sigma R_1^2}{ba(a + 1)} - 3b(a + 1) \tag{5.3}$$

Note that the bigger the differences in the medians of the groups, the larger the value of ΣR_1^2 will be, and so the larger the value of χ^2.

5.7.3 Carrying out the test

The method of carrying out the test is best seen by working through an example.

Example 5.5

In an experiment to determine the effectiveness of quinine and chilli in deterring birds from eating slug pellets, three sets of (non-toxic dummy) slug pellets were tested on birds: controls, with nothing added; ones with added quinine and ones with added chilli. Ten pellets of each type were placed on a bird table and left in a spot popular with garden birds. After half an hour the numbers of pellets of each type left were counted. The pellets were replaced with 10 new ones of each type, and left again for another half hour. The process was repeated 10 times in total. The following results were obtained for the numbers of pellets of each type eaten in each trial. Actual numbers are given first, with the ranks in brackets.

Trial	Control	Plus quinine	Plus chilli
1	4 (2)	5 (3)	2 (1)
2	1 (1.5)	1 (1.5)	4 (3)
3	8 (3)	5 (2)	3 (1)
4	6 (2)	8 (3)	5 (1)
5	2 (1)	4 (2.5)	4 (2.5)
6	4 (1.5)	5 (3)	4 (1.5)
7	2 (1)	4 (2)	5 (3)
8	6 (3)	4 (2)	1 (1)
9	5 (3)	3 (2)	2 (1)
10	9 (3)	2 (1)	5 (2)
Rank sum	21	22	17

It looks as if there were differences, with the birds avoiding chilli more than the other two types of pellet, but were these differences significant?

Step 1: Formulating the null hypothesis

The null hypothesis is that there is no difference in the median scores of the three groups.

Step 2: Calculating the test statistic

Using a calculator

Having worked out the sums of the ranks, we can calculate χ^2 using formula 8.3.

$$\chi^2 = \frac{12 \Sigma R_1^2}{ba(a + 1)} - 3b(a + 1)$$

Where $b = 10$ since birds were tested in 10 trials and $a = 3$ since there were three types of pellet that were being compared.

$$\chi^2 = 12 \times (21^2 + 22^2 + 17^2)/(10 \times 3) \times (3 + 1) - (3 \times 10) \times (3 + 1)$$
$$= 12 \times [441 + 484 + 289]/120 - 120$$
$$= 121.4 - 120 = 1.4$$

Using SPSS

As in repeated measures ANOVA, you must first put all the data into three separate columns, with the results of each trial on the same row. Now click on the **Analyze** menu, move onto the **Nonparametric tests** bar, onto **Legacy Dialogs** and click onto the **k Related Samples** bar. SPSS will come up with the **Tests for Several Related Samples** dialogue box. Put the three variables you want to test into the **Test Variable** box, making sure the **Friedman** test type is ticked. To examine the medians and quartiles click on the **Statistics** tab and in the **Statistics** dialogue box tick **Quartiles.** The completed dialogue boxes and data are shown below.

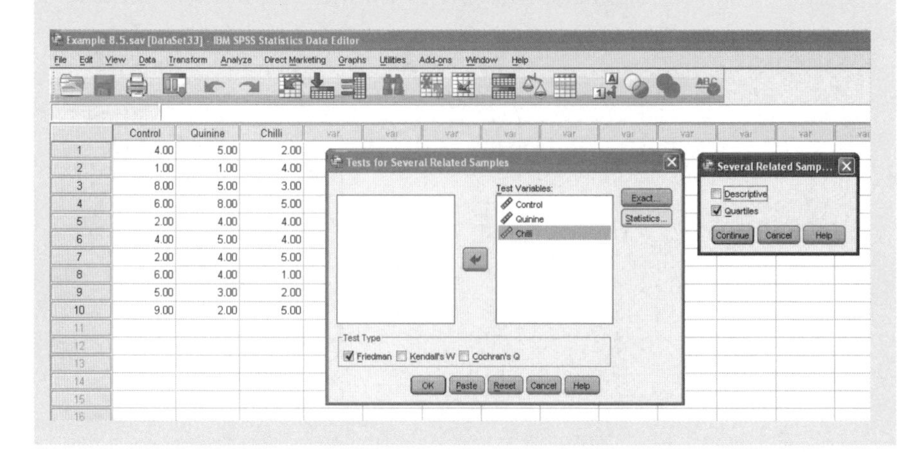

Finally, click on **Continue** and **OK** to run the test. SPSS will come up with the following results.

Descriptive statistics

	N	Percentiles		
		25th	50th (Median)	75th
Control	10	2.0000	4.5000	6.5000
Quinine	10	2.7500	4.0000	5.0000
Chili	10	2.0000	4.0000	5.0000

Friedman test

Ranks

	Mean rank
Control	2.10
Quinine	2.20
Chili	1.70

Test statistics[a]

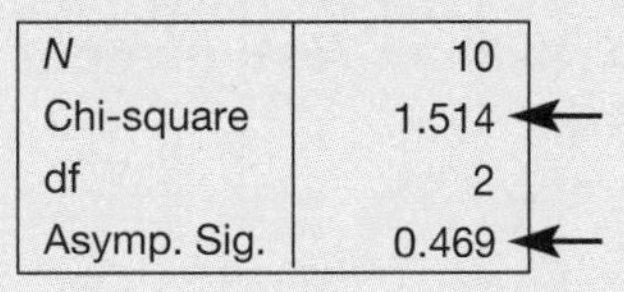

N	10
Chi-square	1.514
df	2
Asymp. Sig.	0.469

[a] Friedman test

SPSS gives the value of chi-square = 1.514.

NB SPSS now has new analysis available for this test, available through the **Independent Samples** bar. I don't recommend this new method, however, as although it carries out the test it doesn't actually present you with any statistics!

Using MINITAB

Putting the data into MINITAB for the Friedman test is rather illogical. First, put all the data into a single column (here called **Number eaten** as for the Kruskall–Wallis test. To distinguish between the three packages, you must create a second, subscript, column (called, say, **treatment**) with one of three values, here 1, 2 and 3. Finally you also have to create a third column with subscripts to identify which trial it was. These are given the numbers 1 to 10. Next, click on the **Stat** menu, move onto the **Nonparametrics** bar, and click on **Friedman**. MINITAB will produce the **Friedman** dialogue box. Put the variable you want to test (here **Number eaten**) into the **Response** box and the first subscript column (here **treatment**) into the **Treatment** box. Finally, put the third column (here **trial**) into the **Blocks** box, to give the data set and completed box shown below.

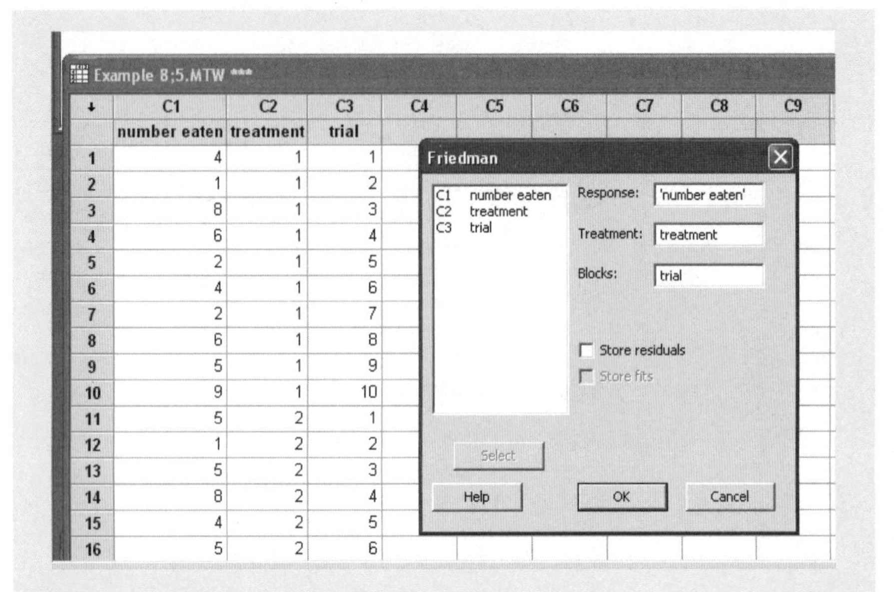

Click on **OK** to run the test.

MINITAB comes up with the following results.

Friedman Test: number eaten versus treatment blocked by trial

```
S = 1.40  DF = 2  P = 0.497
S = 1.51  DF = 2  P = 0.469 (adjusted for ties)
  ↑                  ↑
                              Sum of
treatment    N  Est Median    Ranks
1           10      3.9167     21.0
2           10      4.0833     22.0
3           10      2.7500     17.0

Grand median = 3.5833
```

Once again MINITAB doesn't actually give a value for *K*, but never mind.

Step 3: Calculating the significance probability

You must calculate the probability *P* that the test statistic χ^2 would be **greater than or equal to** χ^2 if the null hypothesis were true.

Using a calculator

You must compare your value of χ^2 with the critical value of the Friedman χ^2 statistic for your values of *a* and *b*. This is given in Table S6 at the end of the book.

Here, looking up in the chi-square distribution for $a = 3$ and $b = 10$ gives a critical value of $\chi^2 = 6.200$.

Using SPSS and MINITAB

SPSS and MINITAB both show that the probability Asymp. Sig. (two-tailed) or P of getting such test statistics is 0.469 (for MINITAB use the lower figure which is adjusted for ties).

Step 4: Deciding whether to reject the null hypothesis

Using a calculator

- If χ^2 is greater than or equal to the critical value, you must reject the null hypothesis. You can say that the medians of the samples are significantly different from each other.
- If χ^2 is less than the critical value, you cannot reject the null hypothesis. You can say that there is no significant difference between the medians of the samples.

Here $\chi^2 = 1.514 < 6.200$.

Using SPSS and MINITAB

- If Asymp. Sig. (two-tailed) or $P \leq 0.05$ you must reject the null hypothesis. Therefore you can say that the medians of the samples are significantly different from each other.
- If Asymp. Sig. (two-tailed) or $P > 0.05$, you have no evidence to reject the null hypothesis. Therefore you can say that there is no significant difference between the medians of the samples.

Here Asymp. Sig. (two-tailed)-$P = 0.469 > 0.05$.

Therefore we have no evidence to reject the null hypothesis. The medians of the samples were not significantly different, so birds did not select significantly different numbers of any of the pellets.

5.7.4 Post hoc tests for the Friedman test

Of course, just as for one-way ANOVA, if there is a significant difference between the groups, you may want to know which groups are different from each other. To do this it is best to carry out separate Wilcoxon matched pairs tests for each test you want to compare, but use the Dunn–Sidak correction, so that the significance probability for each test, rather than being 0.05 is given by the equation $P = (1 - 0.95^{1/k})$ where k is the number of tests.

5.7.5 Presenting the results of the Friedman test

If you are comparing more than two groups, it is best to present your results in the form of a box and whisker plot. For instance the numbers of each sort of pellet is shown in Figure 5.6.

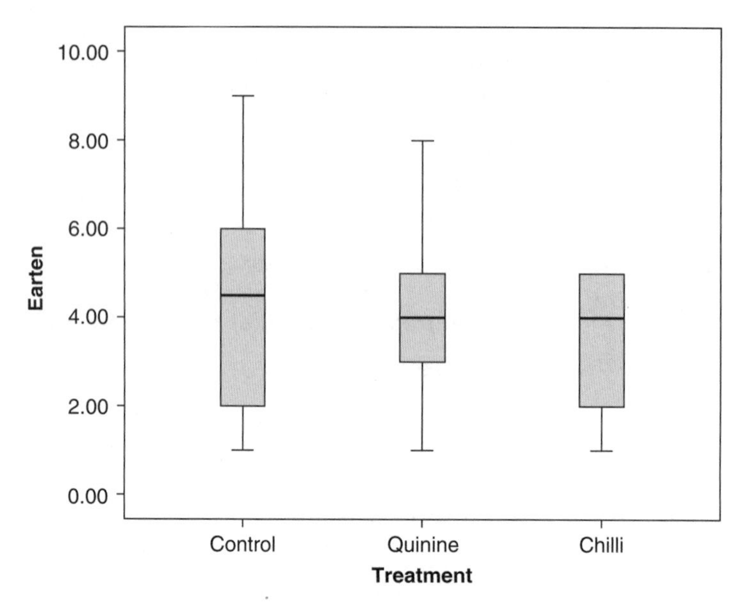

Figure 5.6 Box and whisker plot showing the medians, quartiles and range of the numbers of different flavoured pellets eaten by birds. For all samples $n = 10$.

To show which groups (if any) are different from each other, it is best to use the letter notation as used for Tukey test (Figure 5.3b).

Once again, you should refer to the figure in the text of your results section, saying something like

> The median number of differently flavoured pellets pellets eaten by birds are shown in Figure 5.6. A Friedman test showed that there was no significant difference between the median number eaten.

5.8 Two-way ANOVA

5.8.1 Purpose

To analyse experiments or trials in which you can look at the effect of two factors at once, for instance:

- You might want to examine the effect on corn yield of adding different amounts of nitrate and phosphate.
- You might want to examine the effect on yield of adding different amounts of nitrate to more than one wheat variety.

5.8.2 Rationale

Two-way ANOVA acts in the same way as one-way ANOVA, but with two factors it tests three questions.

1. Are there differences caused by factor 1?
2. Are there differences caused by factor 2?
3. Do the two factors interact with each other? In other words does one factor alter the response of the subjects to the other? Does adding nitrate affect the response of the corn to phosphate? Or does one variety show a greater response to nitrate than the other?

5.8.3 Validity

Like other forms of ANOVA, two-way ANOVA is only valid if the data within each group is normally distributed. It also assumes that the variances of the groups are equal, though this condition is not strict.

5.8.4 Carrying out a test

Carrying out two-way ANOVA tests in SPSS and MINITAB are also fairly long, so once again it is best demonstrated by an example.

Example 5.6

In a field trial to look at the effects of fertilisers, wheat was grown at two different levels of nitrogen and at two different levels of phosphorus. To allow analysis, all possible combinations of nitrogen and phosphorus levels were grown (so there were $2 \times 2 = 4$ combinations in total). The yields (t ha^{-1}) from the experiment are tabulated below.

Treatment									
No nitrate or phosphate Mean = 1.88, s = 0.32, $\overline{SE}$ = 0.105	1.4	1.8	2.1	2.4	1.7	1.9	1.5	2.0	2.1
Added nitrate only Mean = 2.80, s = 0.24, $\overline{SE}$ = 0.082	2.4	2.7	3.1	2.9	2.8	3.0	2.6	3.1	2.6
Added phosphate only Mean = 3.44, s = 0.40, $\overline{SE}$ = 0.132	3.5	3.2	3.7	2.8	4.0	3.2	3.9	3.6	3.1
Added nitrate and phosphate Mean = 6.88, s = 0.61, $\overline{SE}$ = 0.203	7.5	6.4	8.1	6.3	7.2	6.8	6.4	6.7	6.5

Analyse this experiment using **two-way ANOVA** to determine the effects of nitrate and phosphate on yield.

Solution

Step 1: Formulating the null hypotheses

There are three null hypotheses:

1. That nitrate addition had no effect on yield.
2. That phosphate addition had no effect on yield.
3. That there was no interaction between the actions of nitrate and phosphate.

Step 2: Calculating the test statistics

Using SPSS

In SPSS enter all the measurements of yield into a single column and call it, say, yield. Next create two more columns: a second column (called, say, nitrate) with subscripts 0 and 1 for no nitrate and added nitrate, respectively; and a third column (called, say, phosphate) with subscripts 0 and 1 for no phosphate and added phosphate, respectively.

To carry out the test click on the **Analyze** menu, then move to the **General Linear Model** bar and click on **Univariate**. SPSS will come up with the **Univariate** dialogue box. Put yield into the **Dependent Variable** box, and **nitrate** and **phosphate** into the fixed factor box. Next, click on the **Options** tab and click onto the **Descriptive Statistics** box to give you the mean yields etc. for each experimental treatment. The data and completed dialogue boxes are shown below.

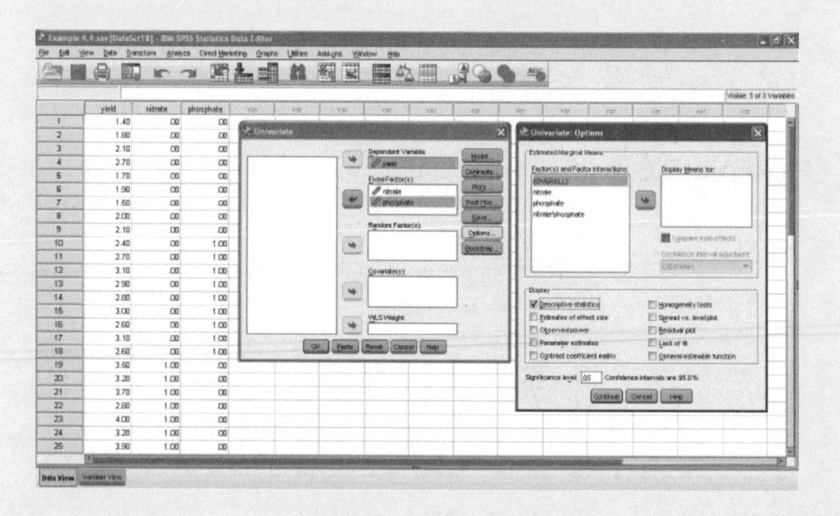

Finally, click on the **Continue** tab to get back to the main dialogue box and click on **OK**. SPSS will produce lots of results, the most important of which are shown below.

Descriptive statistics

Dependent variable: yield

nitrate	phosphate	Mean	Std. deviation	*N*
0.00	0.00	1.9111	0.38550	9
	1.00	3.4444	0.39721	9
	Total	2.6778	0.87552	18
1.00	0.00	2.8000	0.24495	9
	1.00	6.8778	0.60782	9
	Total	4.8389	2.14562	18
Total	0.00	2.3556	0.55436	18
	1.00	5.1611	1.83532	18
	Total	3.7583	1.95176	36

Tests of between-subjects effects

Dependent variable: yield

Source	Type III sum of square	df	Mean square	*F*	Sig.
Corrected model	127.441[a]	3	42.480	230.923	0.000
Intercept	508.503	1	508.503	2764.227	0.000
nitrate	42.034	1	42.034	228.495	0.000 ←
phosphate	70.840	1	70.840	385.089	0.000 ←
nitrate * phosphate	14.567	1	14.567	79.186	0.000 ←
Error	5.887	32	0.184		
Total	641.830	36			
Corrected total	133.328	35			

[a] R squared = 0.956 (Adjusted *R* squared = 0.952).

Just like the one-way ANOVA we have already looked at, two-way ANOVA partitions the variability and variance. However, there will be not two possible causes of variability but four: the effect of nitrate; the effect of phosphate; the **interaction** between the effects of nitrate and phosphate (shown here are nitrate * phosphate) and finally, variation within the groups (here called Error).

These possibilities are here used to produce three F ratios, which test the null hypotheses.

1. The effect of nitrate: $F = 228.5$
2. The effect of phosphate: $F = 385.1$
3. The interaction between nitrate and phosphate: $F = 79.2$

Using MINITAB

In MINITAB enter all the measurements of yield into a single column and call it, say, yield. Next create two more columns: a second column (called, say, nitrate) with subscripts 0 and 1 for no nitrate and added nitrate, respectively; and a third column (called, say, phosphate) with subscripts 0

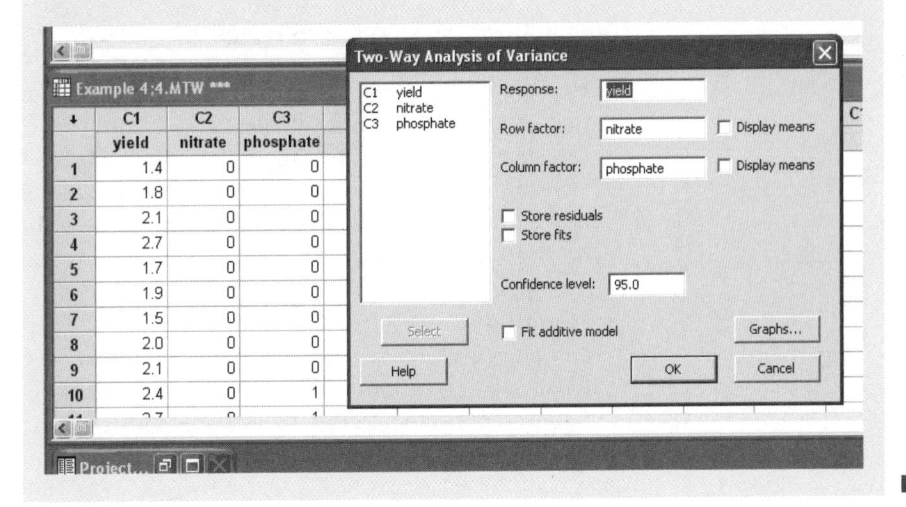

and 1 for no phosphate and added phosphate, respectively. Next, click on the **Stat** menu, move onto the **ANOVA** bar, and click on **Two-Way**. MINITAB will produce the **Two-Way Analysis of Variance** dialogue box. Put the variable you want to test (here **yield**) into the **Response** box and the subscript columns (here **nitrate** and **phosphate**) into the **Row factor** and **Column factor** boxes to give the data set and completed box shown at the bottom of the previous page.

Click on **OK** to run the tests.

MINITAB comes up with the following results.

Two-way ANOVA: yield versus nitrate, phosphate

Source	DF	SS	MS	F	P	
nitrate	1	70.840	70.8403	385.09	0.000	←
phosphate	1	42.034	42.0336	228.50	0.000	←
Interaction	1	14.567	14.5669	79.19	0.000	←
Error	32	5.887	0.1840			
Total	35	133.328				

S = 0.4289   R-Sq = 95.58%   R-Sq(adj) = 95.17%

MINITAB gives the three F ratios, to test the effect of nitrate, phosphate and the interaction between them in the second last column. You can also find out the means and standard deviations of the different groups using the **Basic Statistics** bar.

Step 3: Calculating the significance probability

1. For the effect of nitrate Sig. $= P = 0.000$
2. For the effect of phosphate Sig. $= P = 0.000$
3. For the interaction Sig. $= P = 0.000$

Step 4: Deciding whether to reject the null hypothesis

1. Here Sig. $= P = 0.000 < 0.05$ so we can reject the null hypothesis and say that nitrate has a significant effect on yield. In fact, looking at the descriptive statistics we can see that adding nitrate increases yield by 0.89 t ha^{-1}.
2. Here Sig. $= P = 0.000 < 0.05$ so we can reject the null hypothesis and say that phosphate has a significant effect on yield. In fact, looking at the descriptive statistics we can see that adding phosphate also increases yield by 1.55 t ha^{-1}.
3. Here Sig. $= P = 0.000 < 0.05$ so we can reject the null and say that nitrate and phosphate have a significant interaction. What does this mean? Well, looking at the descriptive statistics we can see that the yield with both nitrate and phosphate is very large. Adding nitrate and phosphate have more effect when added together (they increase yield by 4.97 t ha^{-1} rather than

by just adding the effects of them added singly ($0.89 + 1.55 = 2.44$ t ha^{-1}). In this case they **potentiate** each other's effects. (Though you would also get a significant interaction if they had **inhibited** each other's effects.)

Step 5: Deciding which groups are different

It is also possible to carry out Post hoc tests in SPSS for each of the main effects in two-way ANOVA, just as in one-way ANOVA. In this example, though, since there are only two levels of nitrate and two of phosphate this is not possible. If you have three or more levels, you can just click on the **Post Hoc** tab, put the factors you want to examine into the box and tick the post hoc test you want to perform.

5.8.5 Presenting the results of two-way ANOVA

Like one-way ANOVA, the results of two-way ANOVA tests are best presented in a bar chart. For instance for our results on wheat yield are best presented as in Figure 5.7.

The text should read something like

The results are shown in Figure 5.7. Both nitrate ($F_{1,32} = 385.1$) and phosphate ($F_{1,32} = 228.5$) increased yield significantly and there was also a significant interaction ($F_{1,32} = 79.2$); they potentiated each others effects.

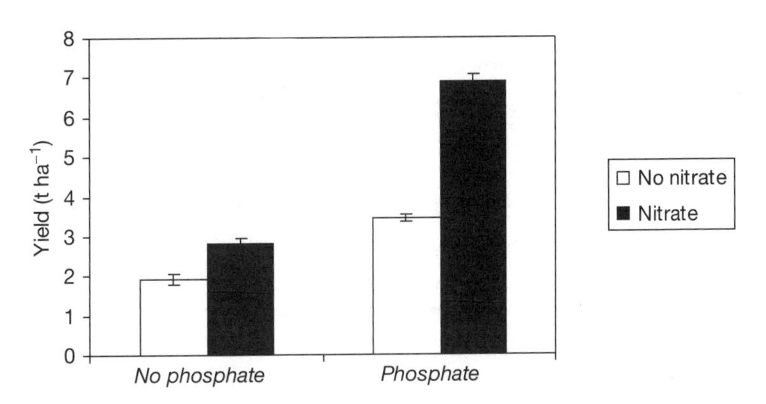

Figure 5.7 The yields of wheat grown in a factorial experiment with or without nitrate and phosphate. Bars show means ± standard error. For all samples $n = 9$.

5.9 The Scheirer–Ray–Hare Test

5.9.1 Purpose

The Scheirer–Ray–Hare test is the non-parametric version of two-way ANOVA, for use when you have ranked or non-normally distributed data. It should be used with caution, however.

5.9.2 Rationale

This test is essentially a two-way extension of the Kruskall–Wallis test, and like two-way ANOVA it tests three questions.

1. Are there differences caused by factor 1?
2. Are there differences caused by factor 2?
3. Do the two factors interact with each other?

5.9.3 Carrying out a test

Carrying out a test is rather tricky, so once again it is best demonstrated by an example.

Example 5.7

In the field trial to look at the effects of fertilisers, the numbers of snails in the different plots were also counted and the results shown below obtained.

No nitrate or phosphate Median = 2	0	1	3	2	1	5	4	8	2
Added nitrate only Median = 6	3	9	5	12	4	9	16	6	1
Added phosphate only Median = 2	2	7	1	2	0	2	0	6	5
Added nitrate and phosphate Median = 6	6	3	12	7	2	17	10	4	5

It look as though areas with higher nitrate had larger numbers of snails, but is this difference significant?

Solution

Step 1: Formulating the null hypotheses

There are three null hypotheses:

1. That nitrate addition had no effect on snail number.
2. That phosphate addition had no effect on snail number.
3. That there was no interaction between the actions of nitrate and phosphate.

Step 2: Calculating the test statistics

Using SPSS

In SPSS enter all the measurements of yield into a single column and call it, say, **number**. Next create two more columns: a second column (called, say, **nitrate**) with subscripts 0 and 1 for no nitrate and added nitrate, respectively; and a third column (called, say, **phosphate**) with subscripts 0 and 1 for no phosphate and added phosphate, respectively.

Before you can carry out the test, you must convert the numbers to ranks. To do this the first thing to do is to sort the data from the lowest number to the highest. To do this click on the **Data** menu and click on **Sort Cases**. SPSS will come up with the **Sort Cases** dialogue box. Put **number** into the **Sort by** box and click on **OK**. Now you can easily create a new **rank** column. In many cases this will just involve typing in an ascending series of numbers, but here there are many cases with equal rank, so it is a bit more tricky (see part of the completed column below). Next, carry out a conventional two-way ANOVA putting **rank** into the **Dependent Variable** box and **nitrate** and **phosphate** into the **fixed factor(s)** box. The data and completed main dialogue box are shown below.

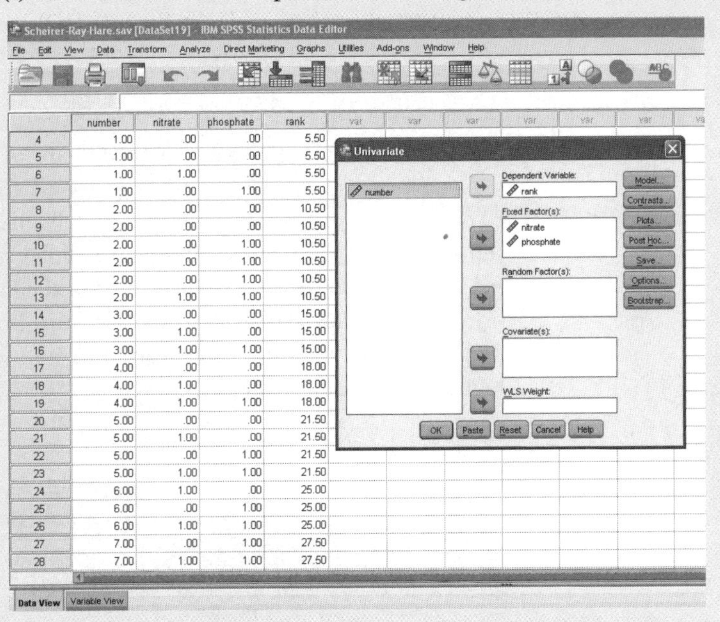

Next, click on **OK** and SPSS will produce the ANOVA table below.

Tests of Between-Subjects Effects

Dependent variable:rank

Source	Type III sum of squares	df	Mean square	*F*	Sig.
Corrected model	1123.722[a]	3	374.574	4.400	0.011
Intercept	12321.000	1	12321.000	144.725	0.000
nitrate	1122.250	1	1122.250	13.182	0.001
phosphate	0.111	1	0.111	0.001	0.971
nitrate * phosphate	1.361	1	1.361	0.016	0.900
Error	2724.278	32	85.134		
Total	16169.000	36			
Corrected total	3848.000	35			

[a] R Squared = 0.292 (Adjusted R Squared = 0.226).

Unfortunately, this is not the end of the matter. You have to calculate the χ^2 test statistics yourself from this table for each factor, where χ^2 = factor sum of squares/total mean square. Here the total mean square = corrected total sum of squares/corrected total df = 3848/35 = 109.94. Therefore

$$\chi^2 \text{ nitrate} = 1122.25/109.94 = 10.21$$
$$\chi^2 \text{ phosphate} = 0.111/109.94 = 0.001$$
$$\chi^2 \text{ interaction} = 1.361/109.94 = 0.013$$

Using MINITAB

In MINITAB enter all the measurements of yield into a single column and call it, say, **number**. Next create two more columns: a second column (called, say, **nitrate**) with subscripts 0 and 1 for no nitrate and added nitrate, respectively; and a third column (called, say, **phosphate**) with subscripts 0 and 1 for no phosphate and added phosphate, respectively. Next, rank the data. To do this go into the **Data** column and click on **Rank**. This brings up the **Rank** dialogue box. Put **number** into the **Rank data in** box and type **rank** into the **Store ranks in** box. The completed data and dialogue boxes are shown below.

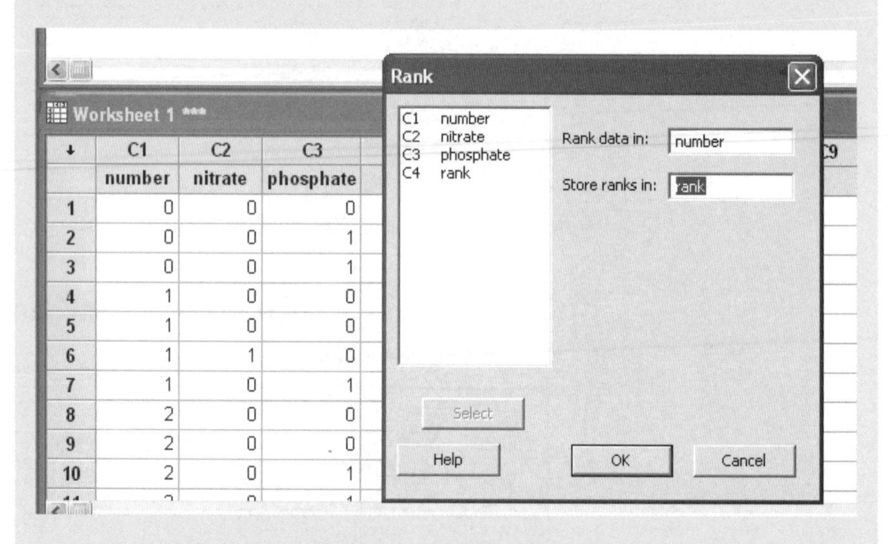

Click on **OK** and MINITAB will produce the ranked data in the new column **rank**. Now carry out a conventional two-way ANOVA. Click on the **Stat** menu, move onto the **ANOVA** bar, and click on **Two-Way**. MINITAB will produce the **Two-Way Analysis of Variance** dialogue box. Put the variable you want to test (here **rank**) into the **Response** box and the subscript columns (here **nitrate** and **phosphate**) into the **Row factor** and **Column factor** boxes. Click on **OK** to run the tests. MINITAB comes up with the following results.

Two-way ANOVA: rank versus nitrate, phosphate

Source	DF	SS	MS	F	P
nitrate	1	1122.25	1122.25	13.18	0.001
phosphate	1	0.11	0.11	0.00	0.971
Interaction	1	1.36	1.36	0.02	0.900
Error	32	2724.28	85.13		
Total	35	3848.00			

S = 9.227   R-Sq = 29.20%   R-Sq(adj) = 22.57%

Unfortunately, this is not the end of the matter. You have to calculate the χ^2 test statistics yourself from this table for each factor, where χ^2 = factor sum of squares/total mean square. Here the total mean square = total SS/ total DF = 3848/35 = 109.94. Therefore

$$\chi^2 \text{ nitrate} = 1122.25/109.94 = 10.21$$
$$\chi^2 \text{ phosphate} = 0.11/109.94 = 0.001$$
$$\chi^2 \text{ interaction} = 1.36/109.94 = 0.013$$

Step 3: Calculating the significance probability

You must calculate the probability P that the test statistics χ^2 would be **greater than or equal to** χ^2 if the null hypothesis were true.

You must compare your value of χ^2 with the critical value of the χ^2 statistic. This is given in Table S3 at the end of the book. The correct number of degrees of freedom are given in the ANOVA tables. For all three factors there is just 1 degree of freedom, so the critical value of chi-square = 3.84.

Step 4: Deciding whether to reject the null hypothesis

- If χ^2 is greater than or equal to the critical value, you must reject the null hypothesis. You can say that the medians of the samples are significantly different from each other.
- If χ^2 is less than the critical value, you cannot reject the null hypothesis. You can say that there is no significant difference between the medians of the samples.

Here χ^2 nitrate = 10.21 > 3.84
χ^2 phosphate = 0.001 < 3.84
χ^2 interaction = 0.013 < 3.84

Therefore we can say that nitrate significantly affects the number of snails, but phosphate has no significant effect and there is no interaction between the factors.

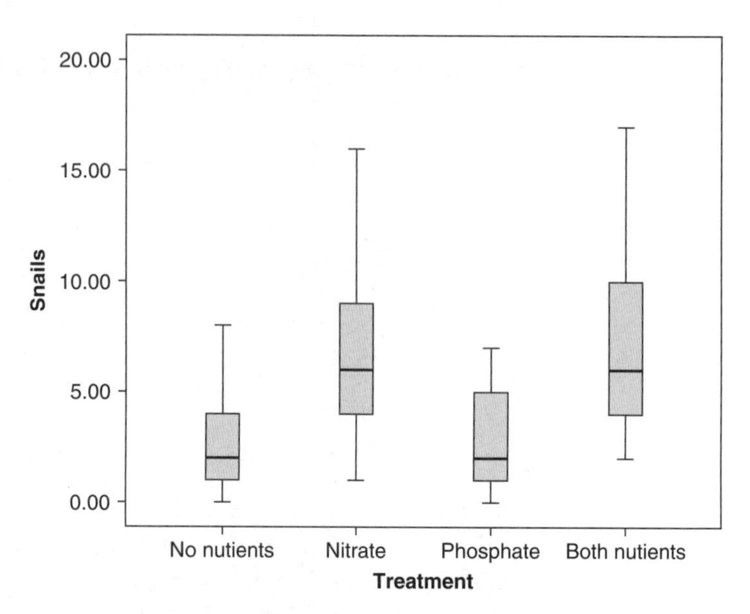

Figure 5.8 Box and whisker plot showing the medians, quartiles and range of the numbers of snails given the different nitrate and phosphate treatments. For all samples $n = 9$.

5.9.4 Presenting the results of the Scheirer–Ray–Hare test

It is best to present your results in the form of a box and whisker plot. For instance the snail data can be presented as shown in Figure 5.8.

Once again, you should refer to the figure in the text of your results section, saying something like

> The median number of snails in the areas given the different treatments is shown in shown in Figure 5.8. A Scheirer–Ray–Hare test showed that there were significantly more snails in areas given high nitrate ($\chi_1^2 = 10.21$, $P < 0.01$), but that neither phosphate nor the interaction between the two factors had a significant effect.

5.10 Nested ANOVA

5.10.1 Purpose

The purpose of a nested ANOVA is to analyse experiments or trials in which you are basically looking at the effect of a single factor, but within each replicate you are taking several measurements. It therefore seems at first glance that there are two factors, the main factor and the replicates, but the second factor is 'nested' within the first. For instance:

- You might want to examine the effect of nitrogen fertilisation on the area of the individual leaves of trees. Here the main factor is nitrogen fertilisation, the replicates are trees, and you are measuring several leaves from each tree.

- You might want to examine the effect of disease on the permeability of individual kidney cells. Here the main factor is disease, the replicates are people and you are measuring several cells from each person.

In these situations, it is tempting just to calculate average values for each replicate (e.g. average leaf area for each tree, or average cell permeability for each person, and analyse the results using a *t* test or one-way ANOVA. However, though this works, you would be missing out information because you would be greatly reducing your sample size, and you would not be able to gain any information about whether the replicates were different from each other. Instead you should carry out a nested ANOVA.

5.10.2 Rationale

Nested ANOVA acts in the same way as one-way ANOVA, but with one factor nested within the other it tests two questions.

1. Are there differences caused by the main factor?
2. Are there differences caused by the replicates?

5.10.3 Validity

Like other forms of ANOVA, nested ANOVA is only valid if the data within each group is normally distributed. It also assumes that the variances of the groups are equal, though this condition is not strict.

5.10.4 Carrying out a test

Carrying out a nested two-way ANOVA tests in SPSS and MINITAB are also fairly long, and so once again it is best demonstrated by an example.

Example 5.8

In an experiment to investigate the effect of keeping flounders in freshwater or salt water on the size of the lice that parasitise them, four fish were kept in freshwater and four in salt water. The lengths (in mm) of the lice living on each fish were measured and the following results were obtained.

Freshwater	Salt water
Fish 1: 1.35, 1.67, 1.57	Fish 5: 1.76, 1.65, 1.69, 1.80
Fish 2: 1.65, 1.55, 1.70, 1.50	Fish 6: 1.56, 1.45
Fish 3: 1.32, 1.28, 1.41, 1.34	Fish 7: 1.89, 1.77, 1.93, 1.87
Fish 4: 1.43, 1.56, 1.60, 1.54, 1.43	Fish 8: 1.76, 1.65, 1.74, 1.83

Note that each fish had different numbers of lice, and there were actually two more lice on fish in freshwater than salt water. This is not a problem! It looks as if the lice in salt water fish are longer but is this difference significant?

Solution

Step 1: Formulating the null hypotheses

There are two null hypotheses:

1. That the water type had no effect on the length of lice.
2. That there was no difference in lice length between fish.

Step 2: Calculating the test statistics

Using SPSS

In SPSS enter all the measurements of lice length into a single column and call it, say, **length**. Next create two more columns: a second column (called, say, **water**) with subscripts 1 and 2 for freshwater and salt water, respectively; and a third column (called, say, **fish**) with subscripts 1 to 8 for the different fish within each water type.

To carry out the test click on the **Analyze** menu, then move to the **General Linear Model** bar and click on **Univariate**. SPSS will come up with the **Univariate** dialogue box. Put **length** into the **Dependent Variable** box, and **water** into the **Fixed Factor(s)** box. Finally, put **fish** into the **Random Factor(s)** box. Next, click on the **Options** tab and click onto the **Descriptive Statistics** box to give you the mean yields etc. for each experimental treatment. The data and completed dialogue boxes are shown below.

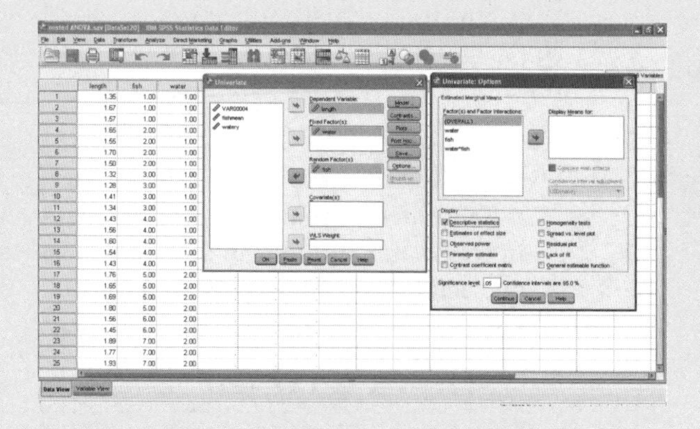

To get SPSS to carry out the correct nested analysis click on **Paste**. This brings up the **Syntax** window shown below.

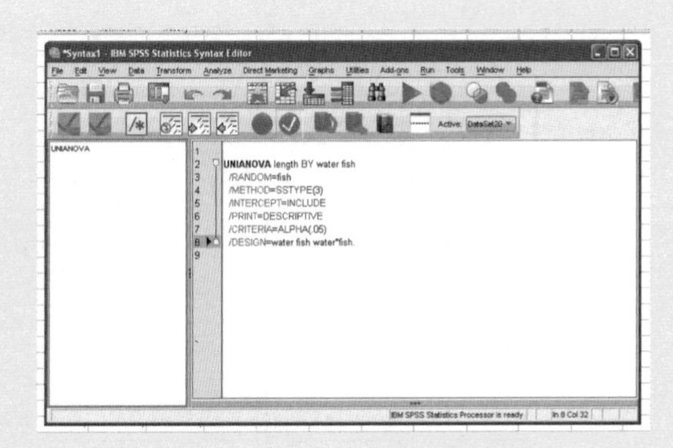

You must change the last line of this to /DESIGN = fish(water) water. This shows that you are looking at two factors: the effect of water type; and the effect of the fish, which are nested within each water type. To run the test, click on the green triangle below **Add-ons** that points to the right. SPSS produces results, the most important bits of which are shown below.

Descriptive statistics

Dependent variable:length

water	fish	Mean	Std. deviation	*N*
1.00	1.00	1.5300	0.16371	3
	2.00	1.6000	0.09129	4
	3.00	1.3375	0.05439	4
	4.00	1.5120	0.07791	5
	Total	1.4938	0.13170	16
2.00	5.00	1.7250	0.06758	4
	6.00	1.5050	0.07778	2
	7.00	1.8650	0.06807	4
	8.00	1.7450	0.07416	4
	Total	1.7393	0.13129	14
Total	1.00	1.5300	0.16371	3
	2.00	1.6000	0.09129	4
	3.00	1.3375	0.05439	4
	4.00	1.5120	0.07791	5
	5.00	1.7250	0.06758	4
	6.00	1.5050	0.07778	2
	7.00	1.8650	0.06807	4
	8.00	1.7450	0.07416	4
	Total	1.6083	0.17950	30

Tests of Between-Subjects Effects

Dependent variable:length

Source		Type III sum of squares	df	Mean square	F	Sig.
Intercept	Hypothesis	71.974	1	71.974	1401.781	0.000
	Error	0.313	6.089	0.051[a]		
fish(water)	Hypothesis	0.322	6	0.054	7.301	0.000
	Error	0.162	22	0.007[b]		
water	Hypothesis	0.324	1	0.324	6.316	0.045
	Error	0.313	6.089	0.051[c]		

[a] 0.949 MS(fish(water)) + .051 MS(Error).

[b] MS(Error).

[c] 0.949 MS(fish(water)) + .051 MS(Error).

The descriptive statistics show that the mean lengths of lice in fresh and salt water are 1.49 and 1.74 mm, respectively. The main table produces two F ratios that are used to test the two null hypotheses.

1. The effect of water: $F = 6.316$
2. The effect of fish: $F = 7.301$.

Using MINITAB

In MINITAB enter all the measurements of length into a single column and call it, say, **length**. Next create two more columns: a second column (called, say, **water**) with subscripts 1 and 2 for fresh and salt water, respectively; and a third column (called, say, **fish**) with subscripts 1 to 8 for the different fish.

Next, click on the **Stat** menu, move onto the **ANOVA** bar, and click on **General Linear Model**. MINITAB will produce the **General Linear Model** dialogue box. Put the variable you want to test (here **length**) into the **Response** box and the two factors (here **water** and **fish**) into the **Model** box. To show that the fish are nested within the water follow **fish** by **water** in brackets. The data set and completed box is shown below.

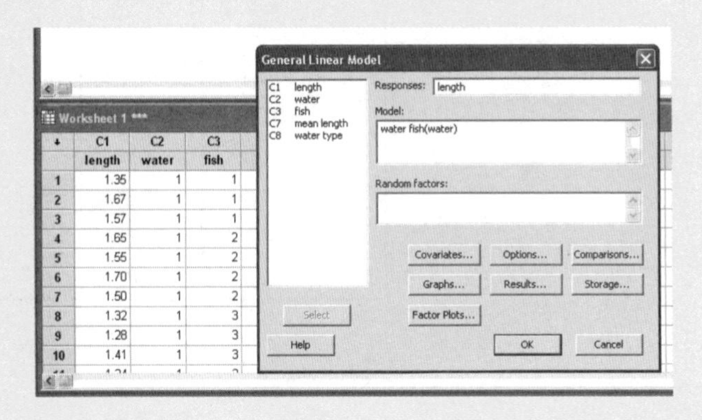

MINITAB comes up with the following results.

General Linear Model: length versus water, fish

```
Factor        Type    Levels  Values
water         fixed        2  1, 2
fish(water)   fixed        8  1, 2, 3, 4, 5, 6, 7, 8

Analysis of Variance for length, using Adjusted SS for Tests

Source        DF   Seq SS   Adj SS   Adj MS      F      P
water          1  0.45015  0.32429  0.32429  44.07  0.000  ←
fish(water)    6  0.32236  0.32236  0.05373   7.30  0.000  ←
Error         22  0.16190  0.16190  0.00736
Total         29  0.93442

S = 0.0857865   R-Sq = 82.67%   R-Sq(adj) = 77.16%

Unusual Observations for length

Obs   length       Fit   SE Fit  Residual  St Resid
  1  1.35000  1.53000  0.04953  -0.18000     -2.57 R

R denotes an observation with a large standardized residual.
```

You can also find out the means and standard deviations of the different groups using the **Basic Statistics** bar.

The main table produces two F ratios that are used to test the two null hypotheses.

1. The effect of water: $F = 44.07$.
2. The effect of fish: $F = 7.30$.

Step 3: Calculating the significance probability

1. For the effect of water Sig. = 0.045; $P = 0.000$.
2. For the effect of fish Sig. = $P = 0.000$.

Step 4: Deciding whether to reject the null hypothesis

1. Here Sig. = 0.045; $P = 0.000 < 0.05$ so we can reject the null hypothesis and say that water type has a significant effect on the length of lice. In fact, looking at the descriptive statistics we can see that lice in salt water fish are longer.

2. Here Sig. = $P = 0.000 < 0.05$ so we can reject the null hypothesis and say that fish have a significant effect on the length of lice; some fish have longer lice than others.

Notes for nested ANOVA's

You may have seen that SPSS and MINITAB calculate rather different values of F and P for the main factor. This is because they use slightly different conventions about how to calculate F. This is rather beyond the scope (or even interest) of anyone except statistics nerds. Note, though, that if we had taken means for the length of lice on each fish and carried out a one-way ANOVA, we would have obtained a significance probability of 0.061. Because of the small sample size we would not have found a significant difference between the lengths of lice in the different water types. This shows the effectiveness of nested ANOVA.

5.10.5 Presenting the results of nested ANOVA

Like one-way ANOVA, the results of two-way ANOVA tests are best presented in a bar chart. For instance our results on the length of fish lice could be presented as shown in Figure 5.9.

Once again, you should refer to the figure in the text of your results section, saying something like

> The mean lengths of the fish lice on the different fish are shown in Figure 5.9. A nested ANOVA showed that lice in sea water were significantly longer than those in freshwater ($F_1 = 44.07$, $P < 0.005$) and lice growing on different fish showed significant differences in length ($F_6 = 7.30$, $P < 0.005$).

Many people also recommend showing the completed ANOVA table, though I am not so keen!

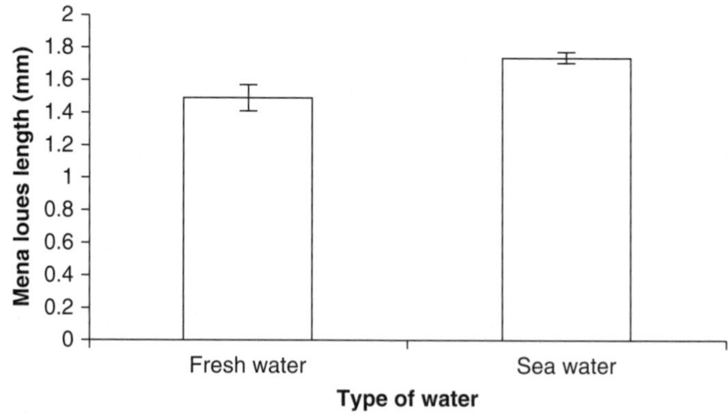

Figure 5.9 Mean (± standard error) lengths of the lice found on fish in fresh water and sea water. For both water types the number of fish = 4.

5.11 Self-assessment problems

Problem 5.1

The levels of calcium-binding protein activity were followed in isolated plant protoplasts following delivery of a heat shock stimulus. Measurements were taken on six samples of protoplasts just before and 1, 2, 4 and 8 hours after the stimulus was applied. The following results were obtained.

Before: 3.2, 2.2, 3.8, 2.8, 2.7, 3.0
1 hour: 3.4, 2.7, 3.2, 4.0, 2.8, 2.9
2 hours: 3.5, 3.7, 4.1, 3.6, 4.7, 3.8
4 hours: 4.5, 4.3, 4.9, 5.1, 3.9, 4.4
8 hours: 3.4, 3.1, 3.6, 2.7, 3.5, 3.2

Investigate the way in which protein activity changes during the time course of the experiment. Carry out a one-way ANOVA and appropriate post hoc tests to determine if any of the apparent changes are significant.

Problem 5.2

Interpret the following ANOVA table. How many groups were being compared? What was the total number of observations? And was there a significant difference between the groups?

	Sum of squares	**df**	**Mean square**	***F***	**Sig.**
Between groups	0.654	0.4	164	1.71	0.35
Within groups	2386	25	0.95		
Total	3040	29			

Problem 5.3

In an experiment to investigate the uptake of aluminium by snails, 20 snails were placed in each of eight tanks of water, each of which had an initial aluminium concentration of 20 mM. The water in each tank was sampled at weekly intervals for five weeks after the start of the experiment and the concentration of aluminium measured. The following results were obtained.

Tank	1	2	3	4	5	6	7	8
Week 1	16.5	14.3	14.6	15.5	13.1	15.2	14.5	13.9
Week 2	12.1	11.2	12.5	10.9	10.5	11.6	13.2	10.5
Week 3	10.9	8.6	10.2	8.7	8.9	9.3	11.0	9.5
Week 4	10.5	7.8	9.6	7.6	6.8	8.0	9.1	8.5
Week 5	10.2	7.4	8.6	7.9	5.7	7.6	8.4	8.2

Carry out a repeated measures ANOVA to test whether aluminium level changes significantly through time. Carry out a post hoc test to determine whether levels *continue* to fall throughout the period.

Problem 5.4

An experiment was carried out to test the effectiveness of three different antibiotics on the germination and growth of bacteria. Bacteria were smeared onto 40 petri dishes: 10 dishes were left as controls, while 10 had antibiotic A, 10 antibiotic B and 10 antibiotic C added. After three days the numbers of bacterial colonies was counted on each dish. The following results were obtained.

Control	0, 6, 9, 1, 2, 8, 3, 5, 2, 0
Antibiotic A	0, 2, 1, 3, 0, 0, 1, 0, 0, 2
Antibiotic B	0, 5, 2, 1, 0, 2, 7, 0, 2, 5
Antibiotic C	6, 1, 5, 2, 0, 1, 0, 7, 0, 0

Carry out a Kruskall–Wallis test to see if the antibiotics had any significant effect on the numbers of bacterial colonies.

Problem 5.5

An experiment was carried out to test the effectiveness over time of an antidepressant drug. Ten patients were asked to assess their mood on a 1 (depressed) to 5 (ecstatic) scale before, one day, one week and one month after taking the drug. The following results were obtained.

Student	1	2	3	4	5	6	7	8	9	10
Before	2	3	2	4	2	1	3	2	1	2
1 day after	4	5	3	4	4	4	3	3	3	3
1 week after	3	3	4	4	3	3	4	3	4	3
1 month after	3	2	3	4	2	2	2	1	2	2

Carry out a Friedman test to determine if the drug had any significant effect on the patients' moods. What pattern emerges of the action of the drug over time and how would you test for it?

Problem 5.6

In a field trial, two different varieties of wheat. Widgeon and Hereward, were grown at three different levels of nitrogen. The following results were obtained.

Widgeon	**Nitrates added (kg m^{-2})**		
	0	**1**	**2**
Yield (t ha^{-1})	4.7	6.4	7.8
	5.3	7.5	7.4
	5.1	6.9	8.3
	6.0	8.1	6.9
	6.5	5.9	6.5
	4.8	7.6	7.2
	5.6	7.1	6.3
	5.8	6.4	7.9
	5.4	8.6	7.7

Hereward	**Nitrates added (kg m^{-2})**		
	0	**1**	**2**
Yield (t ha^{-1})	1.3	6.1	10.8
	2.2	7.2	9.8
	2.1	7.4	11.4
	3.3	8.6	10.6
	1.8	5.7	12.2
	2.4	7.2	9.6
	2.6	6.7	11.1
	2.7	6.9	10.4
	3.1	8.4	10.9

Carry out a two-way ANOVA to answer the following questions.

(a) Which of the three possible effects, variety, nitrogen and interaction, are significant?

(b) Examine the descriptive statistics to work out what these results mean in real terms.

6 Investigating relationships

6.1 Introduction

We saw in Chapter 4 that we can use a paired t test or the Wilcoxon matched pairs test to determine whether two sets of paired measurements are different. For instance, we can test whether students have a different heart rate after drinking coffee compared with before. But we may instead want to know if and how the two sets of measurements are **related**. Do the students who have a higher heart rate before drinking coffee also have a higher heart rate afterwards? Or we might ask other questions. How are the lengths of snakes related to their age? How are the wing areas of birds related to their weight? Or how are the blood pressures of stroke patients related to their heart rate?

This chapter has three sections. First, it shows how to examine data to see whether variables are **related**. Second, it shows how you can use statistical tests to work out whether, despite the inevitable variability, there is a real **linear relationship** between the variables, and if so how to determine what it is. Finally, it describes some of the non-linear ways in which biological variables can be related and shows how data can be transformed to make a linear relationship, the equation of which can be determined statistically.

6.2 Examining data for relationships

scatter plot
A point graph between two variables which allows one to visually determine whether they are associated.

independent variable
A variable in a regression which affects another variable but is not itself affected.

dependent variable
A variable in a regression which is affected by another variable.

The first thing you should do if you feel that two variables might be related is to draw a **scatter plot** of one against the other. This will allow you to see at a glance what is going on. For instance, it is clear from Figure 6.1 that as the age of eggs increases, their mass decreases. But it is important to make sure you plot the graph the correct way round. This depends on how the variables affect each other. One of the variables is called the independent variable; the other variable is called the dependent variable. The independent variable affects, or may affect, the dependent variable but is not itself affected. Plot the **independent variable** along the horizontal axis, often called the x-axis. Plot the **dependent variable** along the vertical axis, often called the y-axis. You would then say you were plotting the dependent variable *against* the independent variable. In

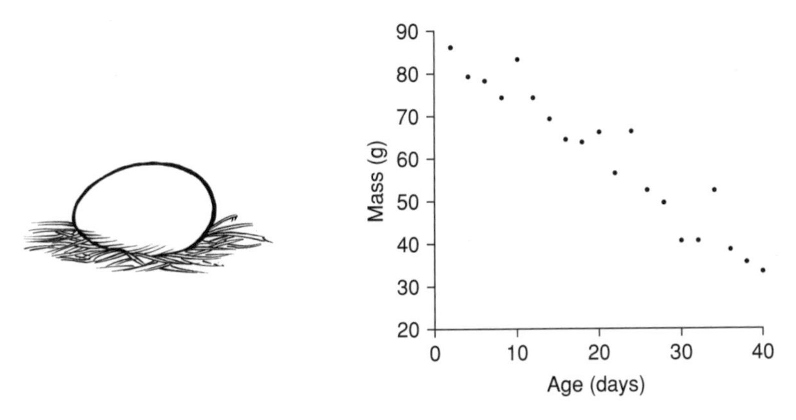

Figure 6.1 The relationship between the age of eggs and their mass. Note that the dependent variable, mass, is plotted along the vertical axis.

Figure 6.1, age is the independent variable and mass is the dependent variable. This is because age can affect an egg's mass, but mass can't affect an egg's age.

Things are not always so clear-cut. It is virtually impossible to tell whether blood pressure would affect heart rate or vice versa. They are probably both affected by a third variable – artery stiffness. In this case, it does not matter so much; however, the one you wish to predict from the relationship (if any) should go on the y axis.

6.3 Examining graphs

Once you have plotted your graph, you should examine it for associations. There are several main ways in which variables can be related:

- There may be no relationship: points are scattered all over the graph paper (Figure 6.2a).
- There may be a positive association (Figure 6.2b): the dependent variable increases as the independent variable increases.
- There may be a negative association (Figure 6.2c): the dependent variable decreases as the independent variable increases.
- There may be a more complex relationship: Figure 6.2d shows a relationship in which the dependent variable rises and falls as the independent variable increases.

6.4 Linear relationships

There are an infinite number of ways in which two variables can be related, most of which are rather complex. Perhaps the simplest relationships to describe linear ones such as that shown in Figure 6.3. In these cases, the dependent variable y is related to the independent variable x by the general equation

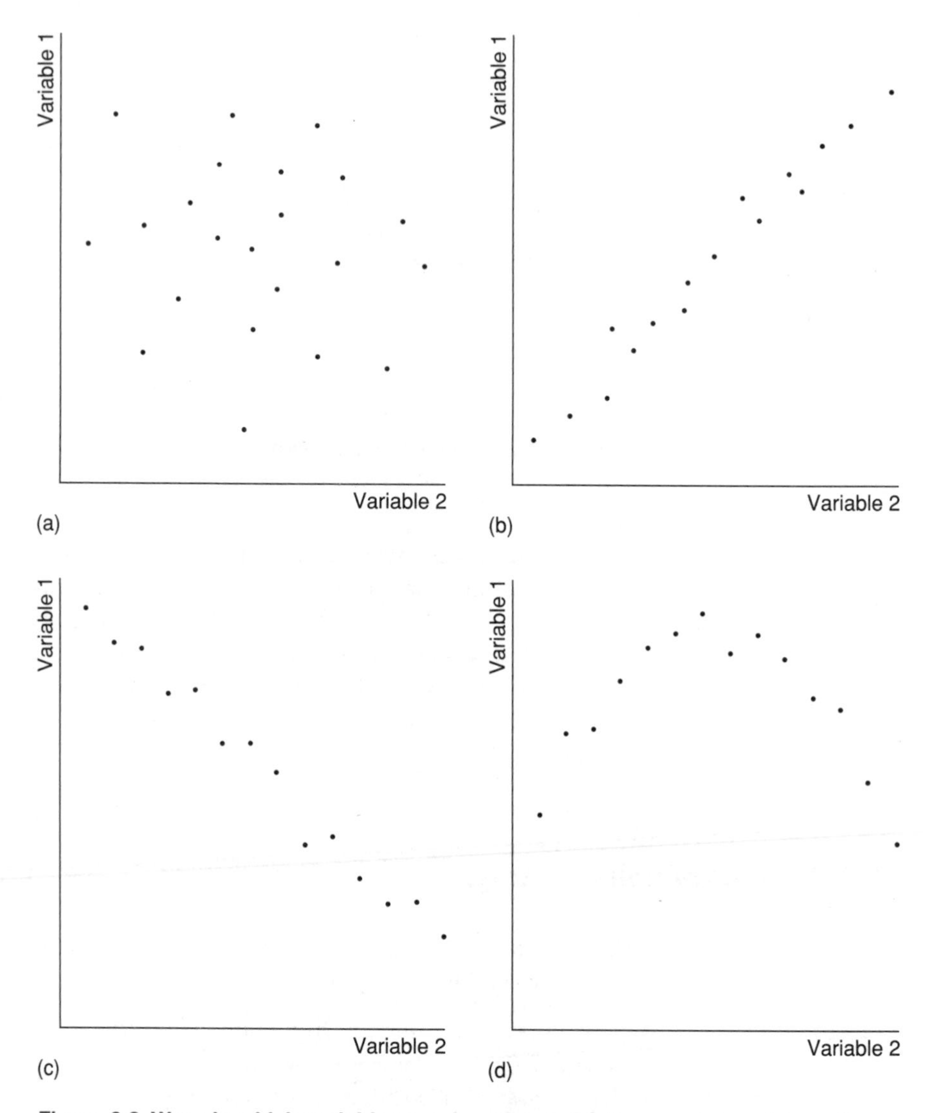

Figure 6.2 Ways in which variables can be related. (a) No association; **(b)** positive association; **(c)** negative association; **(d)** a complex curvilinear association.

$$y = a + bx \quad (6.1)$$

slope
The gradient of a straight line.

intercept
The point where a straight line crosses the y-axis.

where b is the **slope** of the line and a is the constant or **intercept**. The intercept is the value of y where the line crosses the y-axis. Note that this equation is EXACTLY THE SAME as the equation

$$y = mx + c \quad (6.2)$$

which is the form in which many students encounter it at school.

Linear relationships are important because they are by far the easiest to analyse statistically. When biologists test whether two variables are related, they are usually testing whether they are linearly related. Fortunately, linear relationships between variables are surprisingly common in biology. Many other common

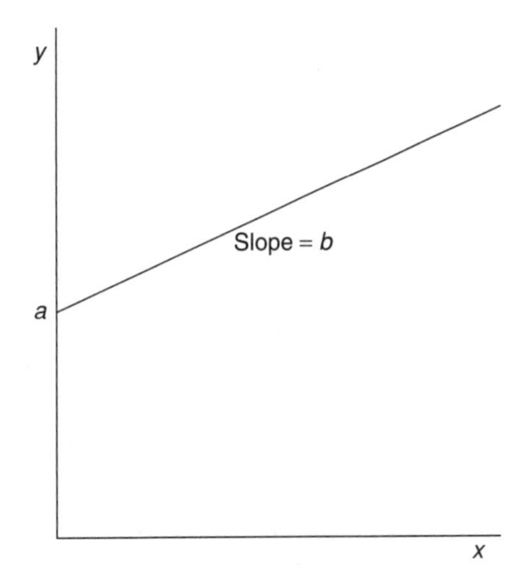

Figure 6.3 A straight line relationship. The straight line $y = a + bx$ has y-intercept a and slope b.

relationships between variables can also be converted into linear relationships by **transforming** the data using logarithms, as we shall see in Section 6.8.

6.5 Statistical tests for linear relationships

The points on your plots will never exactly follow a straight line, or indeed any exact mathematical function, because of the variability that is inherent in biology. There will always be some scatter away from a line. The difficulty in determining whether two measurements are really related is that when you were taking a sample you might have chosen points which followed a straight line even if there were no relationship between the measurement in the population. If there appears only to be a slight association and if there are only a few points, this is quite likely to happen (Figure 6.4a). In contrast it is very unlikely that you would choose large numbers of points all along a straight line just by chance if there was no real relationship (Figure 6.4b). Therefore you have to carry out statistical tests to work out the probability that you could get your apparent relationship by chance. If there is an association, you may also be able to work out what the linear relationship is. There are two main tests for association: correlation and regression.

6.6 Correlation

6.6.1 Purpose

To test whether two sets of paired measurements, neither of which is clearly independent of the other, are linearly associated.

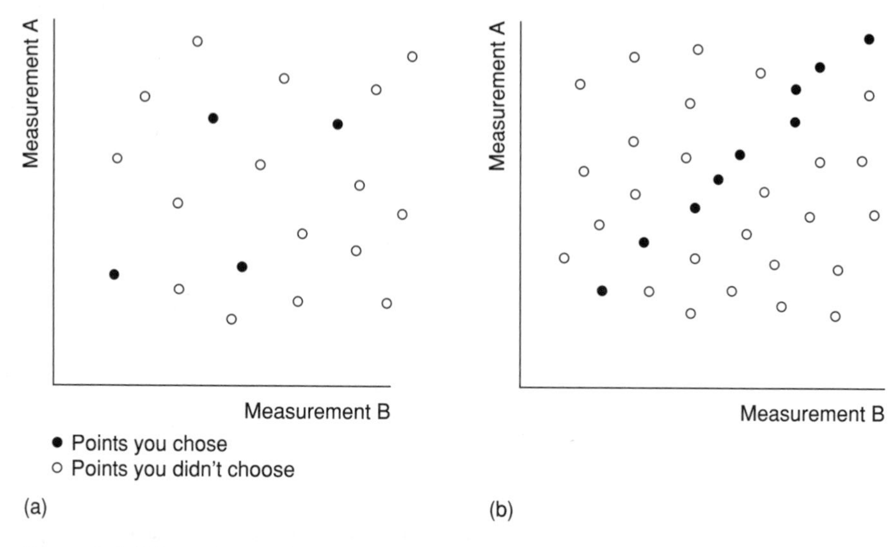

Figure 6.4 Effect of sample size on the likelihood of getting an apparent association. Even if there is no real relationship you might be quite likely to choose a few points which seem to show an association **(a)**. However, if you have a large sample it is very unlikely you would choose points that all fitted onto a straight line **(b)**.

6.6.2 The rationale behind correlation

correlation
A statistical test which determines whether there is linear association between two sets of measurements.

Correlation analysis examines the strength with which two sets of measurements show positive or negative linear association. The basic idea is that if there is positive association, all points will either be above and to the right or below and to the left of the distribution's centre (Figure 6.5a). If there is negative association, all points will be above and to the left or below and to the right of the distribution's centre (Figure 6.5b). If there is no association, the points will be scattered on all sides (Figure 6.5c).

Correlation analysis measures the extent of the association by calculating a single statistic, the **Pearson correlation coefficient** r. It is calculated in three stages.

Stage 1

The first step is to calculate the means of the two sets of measurements, $\bar{x}$ and $\bar{y}$.

Stage 2

The next step is to calculate, for each point, the product of its x and y distances from the mean $(x - \bar{x})(y - \bar{y})$. Note that if both x and y are greater than the mean, this figure will be positive because both $(x - \bar{x})$ and $(y - \bar{y})$ will be positive. It will also be positive if both x and y are smaller than the mean, because both $(x - \bar{x})$ and $(y - \bar{y})$ will be negative and their product will be positive. However, if one is larger than the mean and the other smaller, the product will be negative.

These points are added together to give

$$\text{Sum} = \sum (x - \bar{x})(y - \bar{y})$$

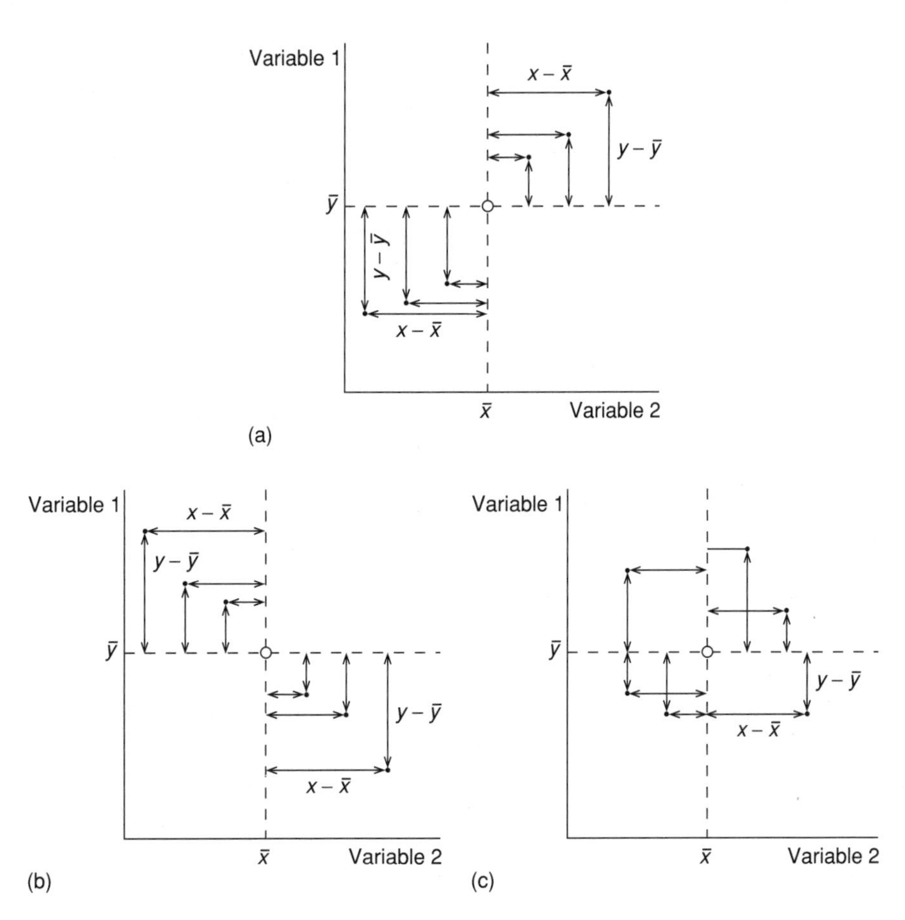

Figure 6.5 Correlation. (a) Positive correlation: $\Sigma(x - \bar{x})(y - \bar{y})$ is large and positive; **(b)** negative correlation: $\Sigma(x - \bar{x})(y - \bar{y})$ is large and negative; **(c)** no correlation: $\Sigma(x - \bar{x})(y - \bar{y})$ is small.

- If there is positive association (Figure 6.5a), with points all either above and to the right or below and to the left of the overall (Figure 6.5b), with points all either above and to the left or below and to the right of the overall mean, the sum will be large and negative.
- If there is no association (Figure 6.5c), points will be on all sides of the overall mean, and the positive and negative numbers will cancel each other out. The sum will therefore be small.

Stage 3

The final stage is to scale the sum obtained in stage 2 by dividing it by the product of the variation within each of the measurements. The correlation coefficient r is therefore given by the formula

$$r = \frac{\Sigma(x - \bar{x})(y - \bar{y})}{[\Sigma(x - \bar{x})^2 \Sigma(y - \bar{y})^2]^{1/2}} \tag{6.3}$$

However, this equation is rather clumsy, and can be simplified to allow it to be calculated somewhat more quickly to give

$$\frac{n\Sigma xy - \Sigma x \, \Sigma y}{\sqrt{[(n\Sigma x^2 - (\Sigma x)^2)(n\Sigma y^2 - (\Sigma y)^2)]}} \tag{6.4}$$

The correlation coefficient can vary from −1 (perfect negative correlation) through 0 (no correlation) up to a value of 1 (perfect positive correlation). The further r is away from zero and the larger the sample size, the less likely it is such a correlation could have occurred by chance if there was no real association between the measurements.

6.6.3 Validity

Both sets of data must be normally distributed.

6.6.4 Carrying out the test

The best way to see how correlation analysis can be performed is to work through an example.

Example 6.1

In an investigation of the cardiovascular health of elderly patients, the heart rate and blood pressure of 30 patients were taken. The following results were obtained. Is there a linear association between the variables?

Patient	Heart rate (min^{-1})	Blood pressure (mmHg)
1	67	179
2	75	197
3	63	175
4	89	209
5	53	164
6	76	180
7	98	212
8	75	187
9	71	189
10	65	176
11	69	167
12	74	186
13	80	198
14	58	170
15	76	187
16	68	175
17	64	169
18	76	190
19	79	176
20	72	168

Patient	Heart rate (min^{-1})	Blood pressure (mmHg)
21	60	158
22	67	160
23	63	167
24	90	221
25	50	149
26	73	180
27	64	168
28	68	162
29	65	168
30	70	157

Solution

Plotting relationship data in SPSS

As well as carrying out statistical tests, SPSS can first be used to graphically examine the data. Simply put the data into two columns and name them, here **heart** and **pressure**. Now go into the **Graphs** menu and click on **Chart Builder.** In the **Chart Builder** dialogue click on **Scatter/Dot** from within the **Choose from** menu and drag the top left simple scatter plot into the big **Chart Preview** box. Finally move **heart** into the **Y Axis** box and **pressure** into the **X Axis** box (actually it doesn't matter which way round in this case). The completed box with the data screen is shown below.

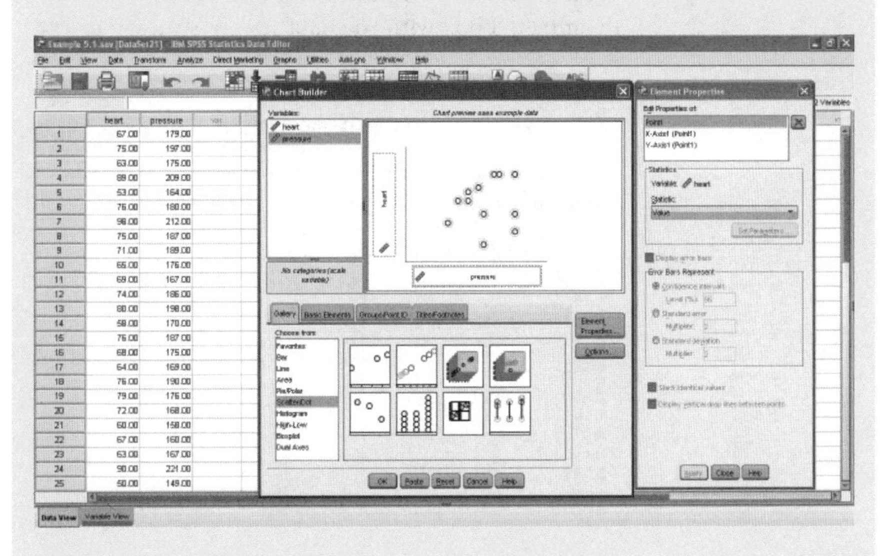

Finally click on **OK** to get SPSS to draw the graph. It will produce a graph like the one shown below.

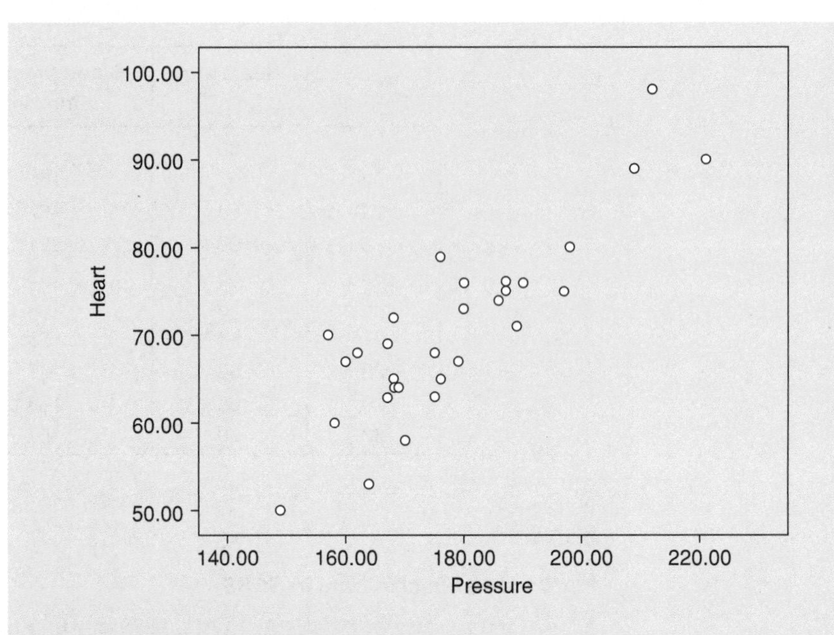

Plotting relationship data in MINITAB

Simply put the data into two columns and name them, here **heart rate** and **pressure**. Now go into the **Graphs** menu and click on **Scatterplot.** In the little **Scatterplots** dialogue box that pops up, click on **Simple** and then **OK.** MINITAB will produce the **Scatterplot - Simple** dialogue box. Move **heart rate** into the **Y variables** box and **pressure** into the **X variables** box. The completed box with the data screen is shown below.

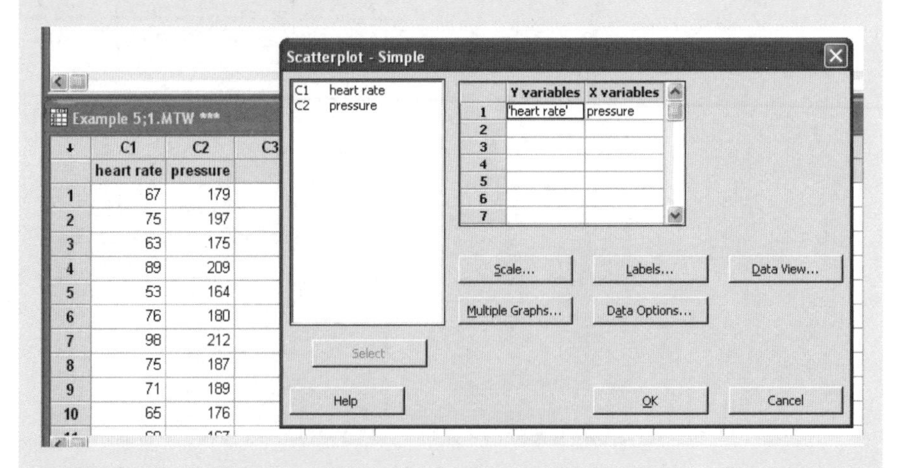

Finally click on **OK** to get MINITAB to draw the graph. It will produce a graph like the one shown below.

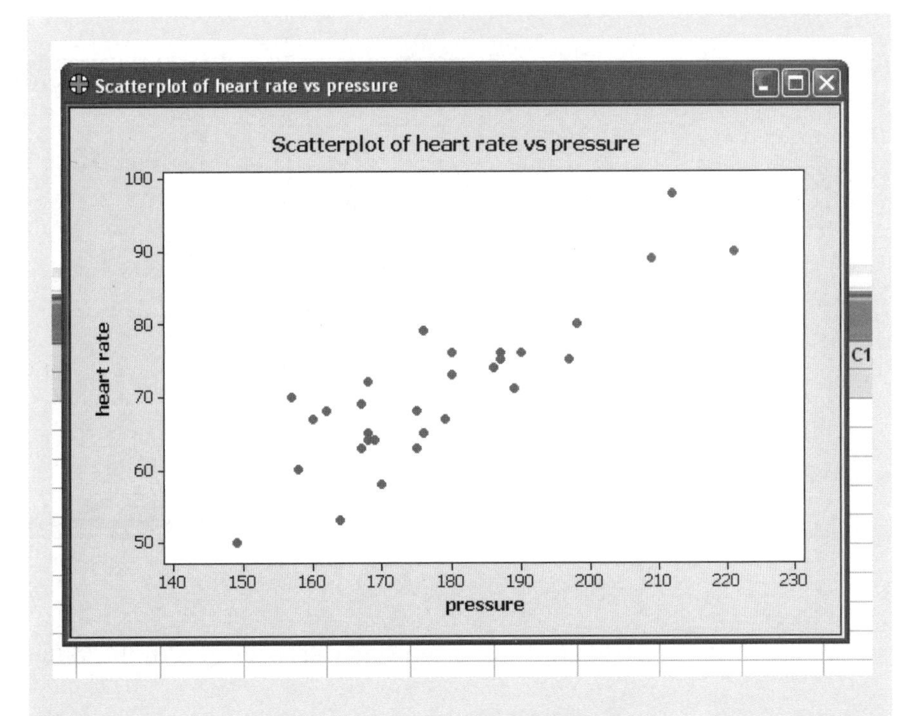

It looks like there is a clear positive correlation, but is it significant? To find out we must carry out a correlation analysis.

Step 1: Formulating the null hypothesis

In correlation, the null hypothesis is that there is no association between the two measurements, i.e. they show random scatter. Here the null hypothesis is that blood pressure and heart rate are not associated.

Step 2: Calculating the test statistic

Using a calculator

Putting the data into equation 6.4, or using the correlation facility of a scientific calculator, gives a correlation coefficient of 0.86.

Using SPSS

Click on the **Analyse** menu, move onto the **Correlation** bar and click on **Bivariate**. SPSS will produce the **Bivariate Correlations** dialogue box. Move **heart** and **pressure** into the **Variables** box and make sure that **Pearson** is ticked. The completed box is shown below.

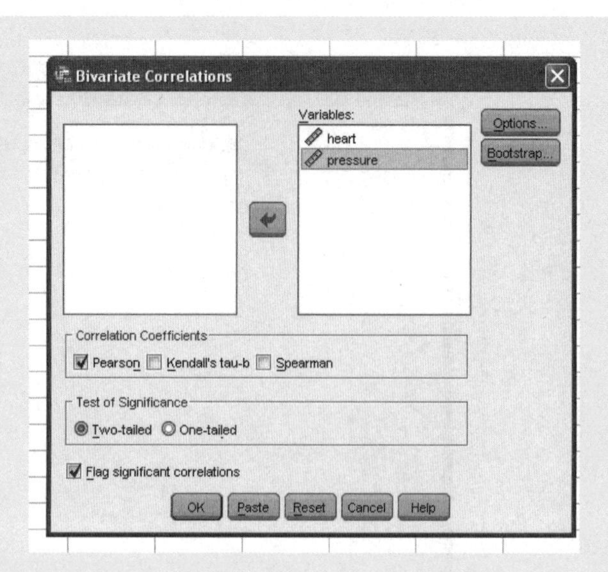

Finally click on **OK**. SPSS will come up with the following results:

Correlations

Correlations

		Heart	Pressure
Heart	Pearson correlation	1	0.860**
	Sig. (two-tailed)	.	0.000
	N	30	30
Pressure	Pearson correlation	0.860**	1
	Sig. (two-tailed)	0.000	.
	N	30	30

** Correlation is significant at the 0.01 level (two-tailed).

Reading from the table the Pearson correlation coefficient $r = 0.860$.

Using MINITAB

Click on the **Stat** menu, move onto the **Basic Statistics** bar and click on **Correlation**. MINIAB will produce the **Correlation** dialogue box. Move **heart rate** and **pressure** into the **Variables** box. The completed box is shown below.

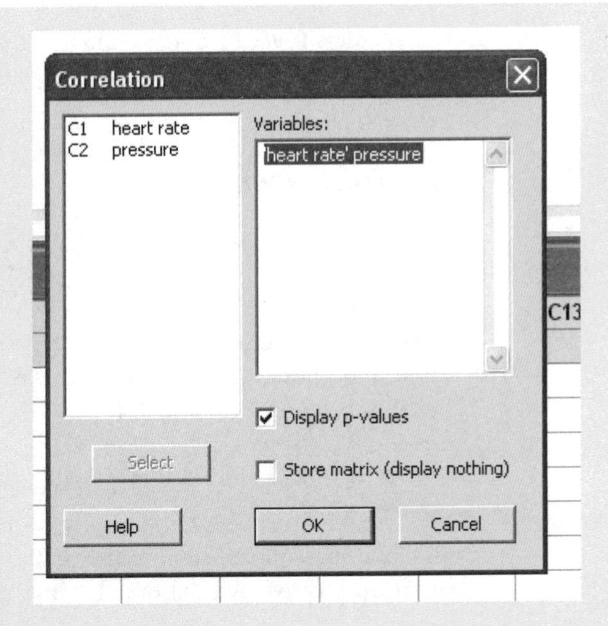

Finally click on **OK**. MINITAB will come up with the following results:

Correlations: heart rate, pressure

```
Pearson correlation of heart rate and pressure = 0.860
P-Value = 0.000
```

MINITAB comes up with a figure for the correlation coefficient of 0.860.

Step 3: Calculating the significance probability

Using a calculator

Critical values of $|r|$ required for P to fall below 0.05, and hence for the association to be significant, are given for a range of degrees of freedom in Table S2 at the end of the book.

You must look up in Table S2 the critical value of r for $(N - 2)$ degrees of freedom, where N is the number of pairs of observations.

Here $N = 30$ so there are 28 degrees of freedom, so $r_{crit} = 0.36$.

Using SPSS and MINITAB

SPSS and MINITAB calculate the significance probability **Sig. (two-tailed)** or ***P*-value** directly. Here it is 0.000.

Step 4: Deciding whether to reject the null hypothesis

Using a calculator

- If $|r|$ is greater than or equal to the critical value, you must reject the null hypothesis. You can say that the two variables show significant correlation.

- If $|r|$ is less than the critical value, you cannot reject the null hypothesis. There is no evidence of a linear association between the two variables.

Here $r = 0.86 > 0.36$.

Using SPSS and MINITAB

- If Sig. (two-tailed) or P-value ≤ 0.05 you must reject the null hypothesis. Therefore you can say that there is a significant association between the variables.
- If Sig. (two-tailed) or P-value > 0.05 you have no evidence to reject the null hypothesis. Therefore you can say that there is no significant association between the variables.

Here Sig. (two-tailed) = P-value = $0.000 < 0.05$.

Therefore we must reject the null hypothesis. We can say that heart rate and blood pressure are significantly correlated. In fact as $r > 0$ they show a significant positive association.

6.6.5 Presenting the results

We saw that before you carry out your correlation analysis you should always examine your data by plotting out a graph. You can also present this in your results. For instance the results of heart rate and blood pressure can be presented as in Figure 6.6.

The graph should be referred to in the text of the results section where the results of the correlation analysis should be given. Here you would say something like

> The relationship between blood pressure and heart rate of the patients is given in Figure 6.6. Correlation analysis showed that there was a significant positive association between heart rate and blood pressure ($r_{28} = 0.860$, $P < 0.001$).

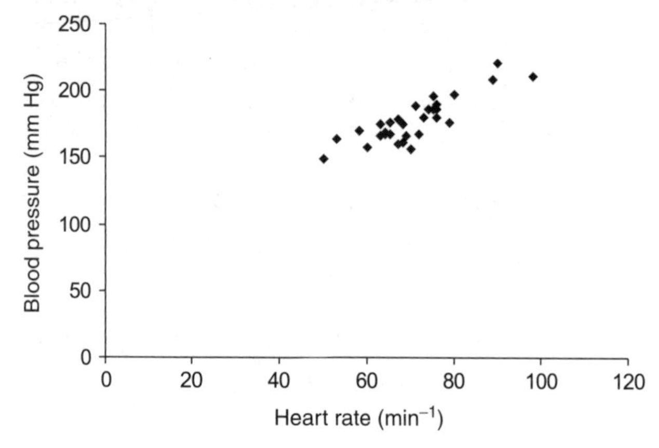

Figure 6.6 Graph showing the relationship between the heart rate and blood pressure of elderly patients.

6.6.6 Uses of the correlation coefficient

Correlation is a useful technique because it tells you whether two measurements are associated, and it can be used even if neither of the variables is independent of the other. However, the results of correlation analysis need to be treated with caution for three reasons:

causal relationship
Relationship between two variables whereby one affects the other but is not itself affected.

- Correlation only finds linear relationships between measurements, so a non-significant correlation does not prove there is no relationship between the variables.
- A significant correlation does not imply a **causal relationship** between the two measurements.
- The size of the correlation coefficient does not reflect the slope of the relationship between the two measurements, it just reflects how close or strong the relationship is. If you want to determine the nature of the linear relationship between two sets of measurements, you need to carry out regression analysis. However, this is only valid if one of the variables is obviously independent of the other and so is plotted along the *x*-axis of your graph.

6.7 Regression

6.7.1 Purpose

To quantify the linear relationship between two sets of paired measurements, one of which is clearly independent of the other. Good examples of independent variables are

- Age or time
- An experimentally manipulated variable, such as temperature or humidity.

6.7.2 Rationale

regression
A statistical test which analyses how one set of measurements is (usually linearly) affected by another.

Regression analysis finds an estimate of the line of best fit $y = a + bx$ through the scattered points on your graph. If you measure the vertical distance of each point from the regression line (Figure 6.7a), the line of best fit is the one which minimises the sum of the squares of the distances.

The estimate of the slope k is actually worked out in a similar way to the correlation coefficient, using the formula

$$\bar{b} = \frac{\Sigma(x - \bar{x})(y - \bar{y})}{\Sigma(x - \bar{x})^2} \tag{6.5}$$

which can be simplified to the following

$$b = \frac{n\Sigma xy - \Sigma x \Sigma y}{n\Sigma x^2 - (\Sigma x)^2} \tag{6.6}$$

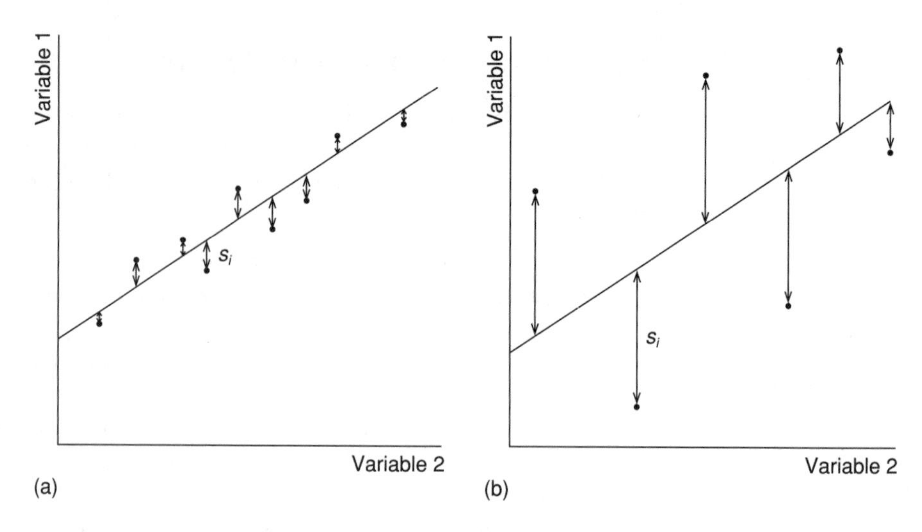

Figure 6.7 Regression. The line of best fit minimises the variability Σs_i^2 from the line. **(a)** Significant regression: Σs_i^2 is low; **(b)** non-significant regression: Σs_i^2 is high.

Since the line of best fit always passes through the means of x and y, $\bar{x}$ and $\bar{y}$, the estimate of the constant a can then be found by substituting them into the equation to give

$$a = \bar{y} - b\bar{x} \tag{6.7}$$

This is all very well, but data with very different degrees of scatter, such as those shown in Figure 6.7, can have identical regression lines. In Figure 6.7a there is clearly a linear relationship. However, in Figure 6.7b there may actually be no relationship between the variables; you might have chosen a sample that suggests there is a relationship just by chance.

In order to test whether there is really a relationship, therefore, you would have to carry out one or more statistical tests. You could do this yourself, but the calculations needed are a bit long and complex, so it is much better now to use a computer package. SPSS not only calculates the regression equation but also performs two statistical tests and gives you the information you need to carry out a whole range of other tests:

standard deviation (σ)
A measure of spread of a group of measurements: the amount by which on average they differ from the mean. The estimate of σ is called S.

***t* tests**
Statistical tests which analyse whether there are differences between measurements on a single population and an expected value, between paired measurements, or between two unpaired sets of measurements

ANOVA
Abbreviation for analysis of variance: a widely used series of tests which can determine whether there are significant differences between groups.

- It works out the **standard deviation** of a and b and uses them to carry out two separate ***t* tests** to determine whether they are significantly different from zero. The data can also be used to calculate 95% confidence intervals for a and b.
- It carries out an **ANOVA** test, which essentially compares the amount of variation explained by the regression line with that due to the scatter of the points away from the regression line. This tells you whether there is a significant slope, e.g. if b is significantly different from zero.
- It also tells you the percentage of the total variation that the regression line explains. This r^2 value is equal to the square of the correlation coefficient.

6.7.3. Validity

Both sets of data must be normally distributed, but you must also be careful to use regression appropriately; there are many cases where it is not valid:

- Simple regression is not valid for data in which there is no independent variable. For example, you should not regress heart rate against blood pressure, because each factor could affect the other. In that case you could, in fact, use **reduced major axis regression**, though debate continues to rage about whether even this approach is really valid. Details of how to carry out this (not too difficult!) analysis can be found in Zar (2000).
- All of your measurements must be independent. Therefore you should not use regression to analyse repeated measures, such as the height of a single plant at different times. If that is the case you will need to use growth analysis, which is a subject in itself.

6.7.4 Carrying out the test

Once again it is best to demonstrate how to carry out a regression analysis by using an example.

Example 6.2

In a survey to investigate the way in which chicken eggs lose weight after they are laid, one egg was collected newly laid every 2 days. Each egg was put into an incubator, and after 40 days all 20 eggs were weighed. The results are tabulated here and plotted in Figure 6.1. Carry out a regression analysis to determine whether age significantly affects egg weight. If there is a relationship, determine what it is.

Age (days)	2	4	6	8	10	12	14	16	18	20	22
Mass (g)	87	80	79	75	84	75	70	65	64	67	57
Age (days)	24	26	28	30	32	34	36	38	40		
Mass (g)	67	53	50	41	41	53	39	36	34		

Solution

Step 1: Formulating the null hypothesis

The null hypothesis is that age has no effect on egg weight. In other words, that the slope of the regression line is zero.

Step 2: Calculating the test statistic

Using SPSS

First put the data into two columns called, say **mass** and **day**. Next, click on the **Analyse** menu, move onto the **Regression** bar and click on **Linear**. SPSS will produce the **Linear Regression** dialogue box. Put the dependent variable (here **mass**) into the **Dependent** box and the independent variable (here **day**) into the **Independent** box. The completed box and the data screen are shown below.

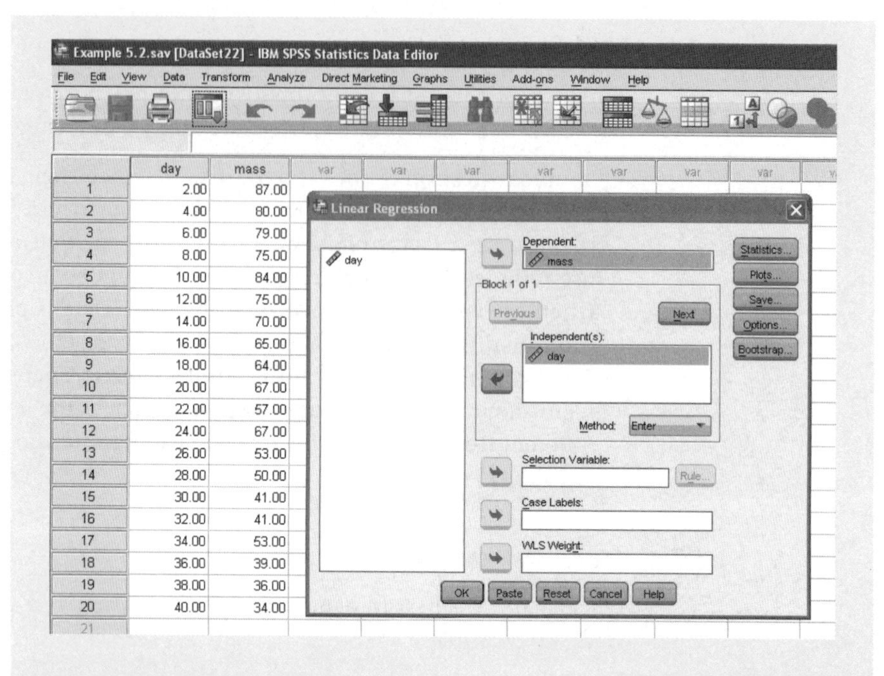

Finally click on **OK**. SPSS will produce masses of output, of which the only useful bit is the following table:

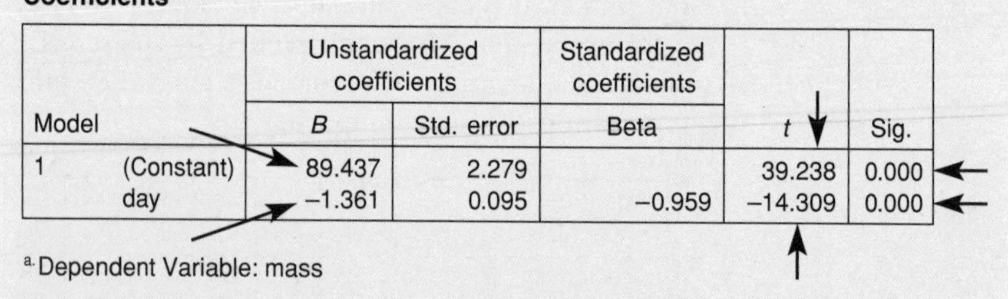

Coefficients[a]

Model		Unstandardized coefficients		Standardized coefficients		
		B	Std. error	Beta	t	Sig.
1	(Constant)	89.437	2.279		39.238	0.000
	day	−1.361	0.095	−0.959	−14.309	0.000

[a] Dependent Variable: mass

This gives you an estimate (*B*) of the Constant and slope (day) from equation 6.1. Here, the line of best fit is the equation

$$\text{Mass} = 89.437-(1.361 \times \text{day}).$$

The slope of the regression equation is −1.36, which appears to be well below zero. But is this difference significant? SPSS has also calculated the standard error of the slope (0.095) and has performed a *t* test to determine whether the slope is significantly different from zero. Here $t = -14.309$.

Using MINITAB

First put the data into two columns called, say **mass** and **day**. Next, click on the **Stat** menu, move onto the **Regression** bar and click on **Regression**. MINITAB will produce the **Regression** dialogue box. Put the dependent variable (here **mass**) into the **Response** box and the independent variable (here **day**) into the **Predictors** box. The completed box, and the data screen are shown below.

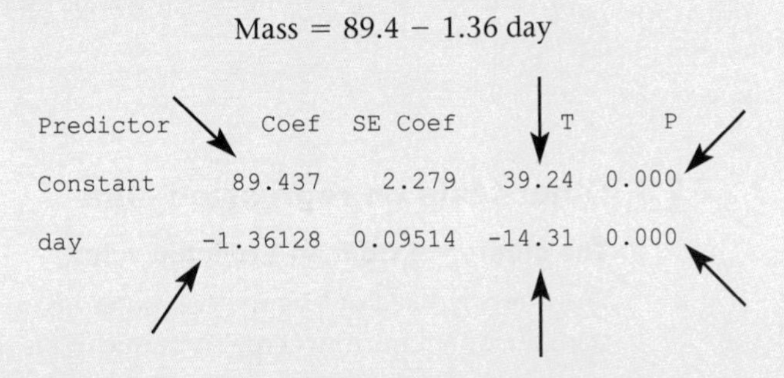

Finally click on **OK**. MINITAB will produce masses of output, of which the most useful bits are the following tables:

The regression equation is

$$\text{Mass} = 89.4 - 1.36 \text{ day}$$

Predictor	Coef	SE Coef	T	P
Constant	89.437	2.279	39.24	0.000
day	-1.36128	0.09514	-14.31	0.000

MINITAB actually gives the regression equation. The slope of the regression equation is −1.36, which appears to be well below zero. But is this difference significant? MINITAB has also calculated the standard error of the slope (0.09514) and has performed a t test to determine whether the slope is significantly different from zero. Here $t = -14.31$.

Step 3: Calculating the significance probability

SPSS and MINITAB have both directly worked out the probability Sig. or P that the slope would be different from 0.

Here Sig. $= P = 0.000$.

Step 4: Deciding whether to reject the null hypothesis

- If Sig. or $P \leq 0.05$ you should reject the null hypothesis. Therefore you can say that the slope is significantly different from zero.
- If Sig. or $P > 0.05$ you have no evidence to reject the null hypothesis. Therefore you can say that the slope is not significantly different from zero.

Here Sig. = $P = 0.000 < 0.05$. Therefore we must reject the null hypothesis. We can say that age has a significant effect on egg weight; in fact older eggs are lighter.

Step 5: Calculating confidence limits

SPSS also calculates confidence intervals for the slope. Simply go into the **Statistics** tab and tick the **Confidence Intervals** box when you are carrying out the test. SPSS will produce the slightly larger final table shown below.

Coefficients[a]

Model		Unstandardized coefficients		Standardized coefficients			95% Confidence interval for B	
		B	Std. error	Beta	*t*	Sig.	Lower bound	Upper bound
1	(Constant)	89.437	2.279		39.238	0.000	84.648	94.226
	day	−1.361	0.095	−0.959	−14.309	0.000	−1.561	−1.161

[a] Dependent variable: mass

This shows that the confidence intervals for the slope of the line are between −1.561 and −1.161.

6.7.5 Other tests on regression data

The difference from an expected value

The *t* tests worked out by the computer investigate whether the slope and constant are different from zero. The value of *t* is simply given by the expression

$$t = \frac{\text{Observed value} - 0}{\text{Standard deviation}} \tag{6.8}$$

However, it is also possible from the computer output to carry out a whole range of **one-sample *t* tests** to determine whether the slope or constant is different from any expected value. Then *t* is simply given by the expression

$$t = \frac{\text{Observed value} - \text{Expected value}}{\text{Standard deviation}} \tag{6.9}$$

Just like equation 4.1, and you can carry out the *t* test for $N - 2$ degrees of freedom just as the computer did to determine whether the slope or constant was different from zero.

Example 6.3

From the egg weight data in Example 6.2, we want to determine whether the initial egg weight was significantly different from 90 g, which is the mean figure for the general population. In other words, we must test whether the intercept (or Constant as SPSS calls it) is different from 90.

Solution

Step 1: Formulating the null hypothesis

The null hypothesis is that the constant is equal to 90.

Step 2: Calculating the test statistic

The necessary data can be extracted from the SPSS or MINITAB output in Example 6.2 (see p. 147). This shows that the estimate of the intercept = 89.437 and its standard deviation = 2.279. Therefore if the expected value = 90, the test statistic is

$$t = \frac{89.437 - 90}{2.279} = -0.247$$

Step 3: Calculating the significance probability

Here $|t|$ must be compared with the critical value for $20 - 2 = 18$ degrees of freedom. This is 2.101.

Step 4: Deciding whether to reject the null hypothesis

We have $|t| = 0.247 < 2.101$. Hence there is no evidence to reject the null hypothesis. We can say that initial egg mass is not significantly different from 90 g.

The difference between two regression lines

If you have calculated two regression lines and want to test whether their slopes or intercepts are *different from each other* you can also carry out **two-sample *t* tests**, using equations 3.5 and 3.6 to give, for instance

$$t = \frac{\text{Slope 1} - \text{Slope 2}}{\sqrt{[(SE_{\text{slope 1}})^{2} + (SE_{\text{slope2}})^{2}]}} \qquad (6.10)$$

Where there are $N + M - 4$ degrees of freedom, where N and M are the two sample sizes.

6.7.6 Presenting the results

If you have carried out regression analysis there is little point in presenting the graph, because the relationship between the two variables is summarised in the regression equation. Instead you can simply present the regression equation along with the results of any of the *t* tests that are relevant. For instance for the egg weight data you might write

Regression analysis showed that the mass of eggs declined significantly with their age ($t_{18} = -14.3$, $P < 0.001$), the line being of best fit being given by the equation

$$\text{Mass (g)} = 89.44 - (1.361 \times \text{age}).$$

Analysis also showed that the mass of eggs at day 0 was not significantly different from 90 g ($t_{18} = -0.247$ $P > 0.05$).

6.8 Studying common non-linear relationships

As we have seen, not all relationships between variables are linear. There are in fact two particularly common non-linear ways in which measurements in biology may be related. Fortunately, as we shall see, these relationships can be changed into linear relationships by transforming one or both of the variables, allowing them to be quantified using regression analysis.

6.8.1 Scaling and power relationships

power relationships
A relationship which follows the general equation $y = ax^b$.

If you examine organisms of different size, many of their characteristics scale according to **power relationships.** If an organism changes in size by a given ratio, some characteristic will increase or decrease by the square, cube or some other power of that ratio. For instance, the mass of unicellular algae would be expected to rise with the cube of their diameter; and the metabolic rate of mammals rises with mass to the power of around 0.75. Other physical processes are also related in this way. The lift produced by a bird's wings should rise with the square of the flight speed.

In these sorts of relationships, the dependent variable y is related to the independent variable x by the general equation

$$y = ax^b \tag{6.11}$$

Looking at the curves produced by this sort of relationship (Figure 6.8a), it is very difficult to determine the values of a and b. However, it is possible, by using some clever mathematical tricks, to produce a straight line graph from

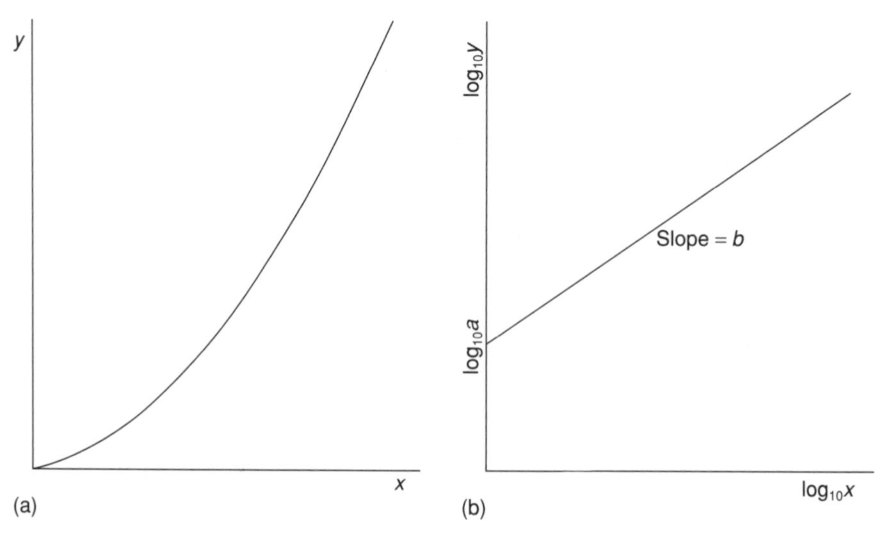

Figure 6.8 How to describe a power relationship. The curvilinear relationship $y = ax^b$ **(a)** can be converted to a straight line **(b)** by taking logarithms ($\log_{10}$) of both x and y. The graph has y-intercept $\log_{10} a$ and slope b.

which a and b can be easily calculated. The first thing to do is to take logarithms of both sides of the equation. We have $y = ax^b$, so

$$\log_{10} y = \log_{10}(ax^b) \quad (6.12)$$

Now logarithms have two important properties:

$$\log_{10}(c \times d) = \log_{10} c + \log_{10} d \quad (6.13)$$

$$\log_{10}(c^d) = d \times \log_{10} c \quad (6.14)$$

Using these properties we can rearrange the equation to show that

$$\log_{10} y = \log_{10} a + b \log_{10} x \quad (6.15)$$

Therefore plotting $\log_{10} y$ against $\log_{10} x$ (Figure 6.8b) will produce a straight line with slope b and intercept $\log_{10} a$.

6.8.2 Exponential growth and decay

exponential relationship A relationship which follows the general equation $y = ae^{bx}$. If $b > 0$ this is exponential growth; if $b < 0$ this is exponential decay.

Other biological phenomena have an **exponential relationship** with time. In these cases, when a given period of time elapses, some characteristic increases or decreases by a certain ratio. For instance, bacterial colonies demonstrate exponential growth, doubling in number every few hours. In contrast, radioactivity shows exponential decay, halving over a given period. Other physical processes are also related in this way. Rates of reaction, indeed the metabolic rates of many whole organisms, increase exponentially with temperature.

In these sorts of relationship the dependent variable y can be related to the independent variable x by the general equation

$$y = ae^{bx} \quad (6.16)$$

where e is the base of natural logarithms ($e = 2.718$). Looking at the curve produced by this sort of relationship (Figure 6.9a), it is very difficult to determine

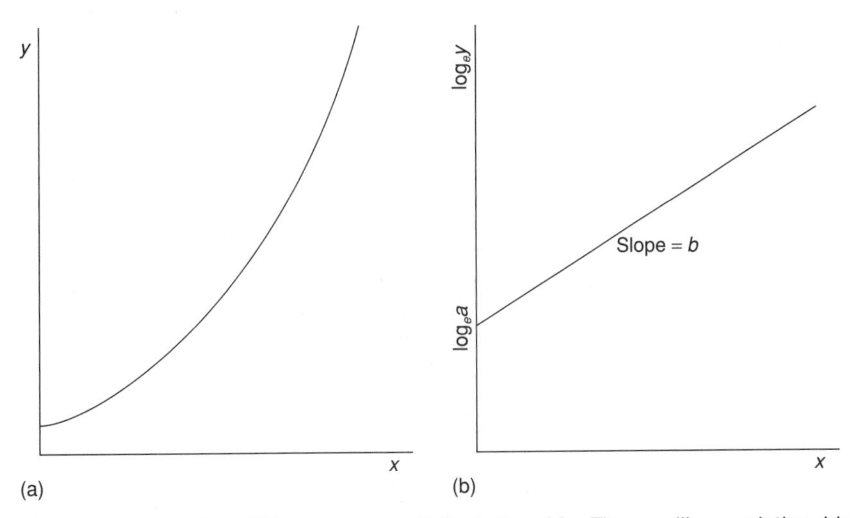

Figure 6.9 How to describe an exponential relationship. The curvilinear relationship $y = ae^{bx}$ **(a)** can be converted to a straight line **(b)** by taking natural logarithms ($\log_e$) of y. The graph has y-intercept $\log_e a$ and slope b.

the values of a and b, just as it was with power relationships. However, we can again use some clever mathematical tricks to produce a straight line graph. As before, the first thing to do is to take logarithms of both sides of the equation. Therefore

$$\log_e y = \log_e(ae^{bx}) \tag{6.17}$$

and rearranging

$$\log_e y = \log_e a + bx \tag{6.18}$$

Therefore plotting $\log_e y$ against x (Figure 6.9b) will produce a straight line with slope b and y-intercept $\log_e a$.

Example 6.4

An investigation was carried out into the scaling of heads in worker army ants. Body length and jaw width were measured in 20 workers of contrasting size. The following results were obtained.

Length (mm)	3.2	3.6	4.2	4.3	4.6	5.0	5.2	5.3	5.5	5.5
Jaw width (mm)	0.23	0.29	0.32	0.38	0.45	0.44	0.55	0.43	0.60	0.58
Length (mm)	5.7	6.2	6.6	6.9	7.4	7.6	8.5	9.2	9.7	9.9
Jaw width (mm)	0.62	0.73	0.74	0.88	0.83	0.93	1.03	1.15	1.09	1.25

It was suggested that these ants showed allometry, the jaws of larger ants being relatively wider than those of smaller ants. It certainly looks that way as the largest ants are around three times as long as the smallest ones but have heads that are around four to five times wider. To investigate whether there is a significant change in proportions, the body length and jaw width must first both be log transformed using SPSS or MINITAB (to see how to transform data see Section 3.5). This gives data which, when plotted in SPSS, gives the following graph.

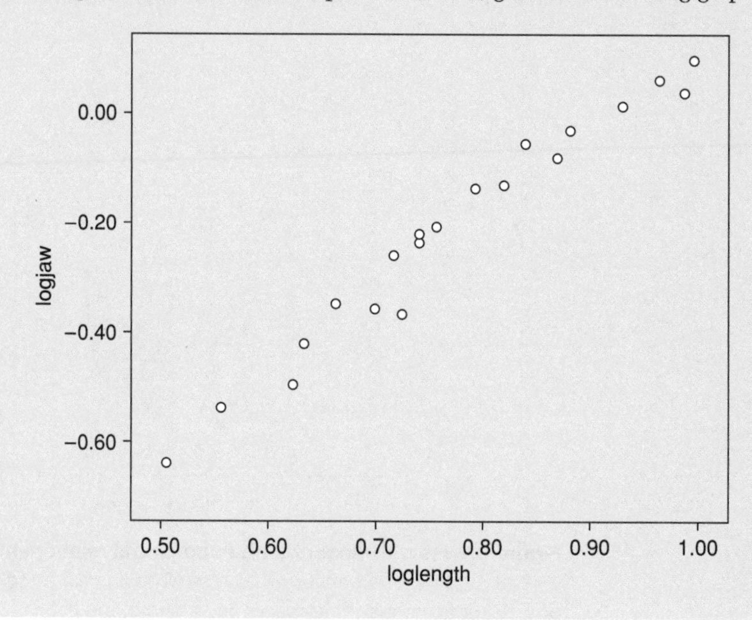

If ants scaled isometrically, head width would be directly proportional to length, so the slope of the log/log graph would equal 1. Therefore, we need to determine whether the slope of this graph is significantly different from 1.

Solution

Step 1: Formulating the null hypothesis

The null hypothesis is that the slope of the line is equal to 1.

Step 2: Calculating the test statistic

Performing a regression analysis in SPSS of logjaw vs loglength gives the following results.

Coefficients[a]

Model		Unstandardized coefficients		Standardized coefficients		
		B	Std. error	Beta	*t*	Sig.
1	(Constant)	−1.362	.052		−26.294	0.000
	loglength	1.485	.066	0.983	22.471	0.000

[a] Dependent variable: logjaw

And in MINITAB gives

Regression Analysis: Log jaw vs Log length

```
The regression equation is

Log Jaw = - 1.36 + 1.49 Log Length

Predictor       Coef   SE Coef       T      P

Constant    -1.36194   0.05180  -26.29  0.000

Log Length   1.48542   0.06610   22.47  0.000
```

The regression line of best fit is given, therefore, as

$$\text{Logjaw} = 1.485\ (\text{loglength}) - 1.362.$$

This means that the equation relating jaw width to length is

$$\text{Jaw width} = 10^{-1.362} \times \text{length}^{1.485}$$

$$\text{Jaw width} = 0.0435\ \text{length}^{1.485}$$

It looks as if the exponent (1.485) *is* greater than 1, but to work it whether there is a significant difference we must carry out a *t* test using the statistics SPSS has calculated.

$$t = \frac{\text{Slope} - \text{Expected slope}}{\text{Standard error of slope}}$$

$$t = (1.485-1)/0.066$$
$$t = 7.35$$

Step 3: Calculating the significance probability

Here $|t|$ must be compared with the critical value for $20 - 2 = 18$ degrees of freedom. This is 2.101.

Step 4: Deciding whether to reject the null hypothesis

We have $|t| = 7.35 > 2.101$. Hence there is good evidence to reject the null hypothesis. In fact we can say that the slope is significantly greater than 1. This means that the ants do show positive allometry, larger ants having relatively wider jaws than smaller ones.

6.9 Dealing with non-normally distributed data: rank correlation

rank
Numerical order of a data point.

If you have **rank** data or if your data is not normally distributed, it is not valid to use either correlation or regression. In these cases, **rank correlation** can be used to see whether there is a relationship between the ranks of the data.

6.9.1 Purpose

To test whether the ranks of two sets of paired measurements are linearly associated.

6.9.2 The rationale behind rank correlation

There are several measurements of rank correlation, but we will examine the one which is most commonly used in biology, **Spearman rank correlation**, given the name r_s or ρ (rho). This works in much the same way as **Pearson correlation**, except that it uses the *ranks* of the observations, rather than the observations themselves.

The formula for calculating ρ is:

$$\rho = 1 - \frac{6\Sigma d^2}{(n^3-n)} \qquad (6.19)$$

Where n is the sample number and d is the difference between ranks for each point. Note that the higher the correlation, the smaller the differences between the ranks, so the higher ρ will be.

6.9.3 Carrying out the test

The method of carrying out the test is best seen by working through an example.

Example 6.5

In a field survey which was investigating if there was any relationship between the density of tadpoles and their dragonfly larvae predators, 12 ponds were sampled. The following results were found.

Dragonfly density	3,	6,	5,	1,	1,	4,	9,	8,	2,	5,	7,	11
Tadpole density	86,	46,	39,	15,	41,	52,	100,	63,	60,	30,	72,	71

A graph generated by SPSS gave the following results, suggesting a positive association. But was this significant?

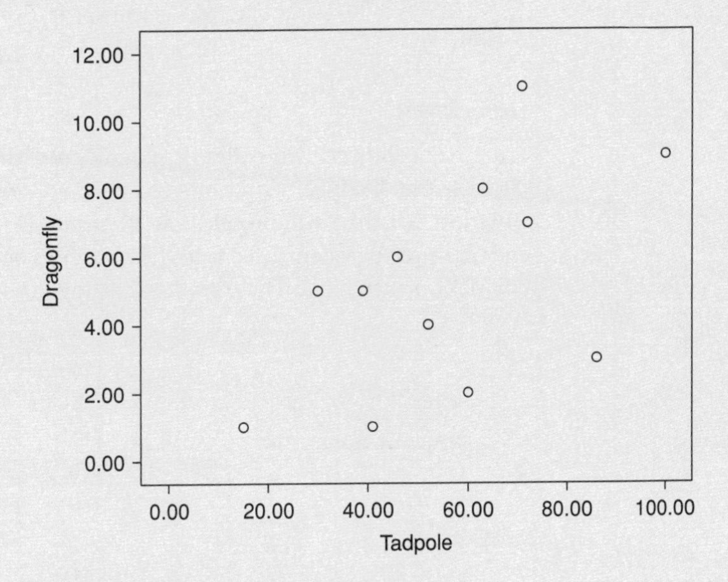

Step 1: Formulating the null hypothesis

The null hypothesis is that there is no association between the density of tadpole and dragonfly larvae.

Step 2: Calculating the test statistic

Using a calculator

You must first work out the ranks of all the observations.

Dragonflies	Rank	Tadpoles	Rank	d	d^2
3	4	86	11	−7	49
6	8	46	5	3	9
5	6.5	39	3	3.5	12.25
1	1.5	15	1	0.5	0.25
1	1.5	41	4	−2.5	6.25
4	5	52	6	−1	1
9	11	100	12	−1	1
8	10	63	8	2	4
2	3	60	7	−4	16
5	6.5	30	2	4.5	20.25

Dragonflies	Rank	Tadpoles	Rank	*d*	d^2
7	9	72	10	−1	1
11	12	71	9	3	9
					$\Sigma d^2 = 129$

$$\text{Since } \rho = 1 - \frac{6\Sigma d^2}{(n^3 - n)}$$

$$\rho = 1-[(6 \times 129)/(123-12)]$$
$$= 1-(774/1716) = 1-0.453 = 0.547$$

Using SPSS

Enter the data into two columns, named, say, **dragonfly** and **tadpole** Next, click on the **Analyze** menu, move onto the **Correlate** bar, then click on the **Bivariate** bar. SPSS will produce the **Bivariate Correlations** dialogue box. Put the columns to be compared into the **Variables** box and tick on the **Spearman** correlation coefficient. The completed dialogue box and data are shown below.

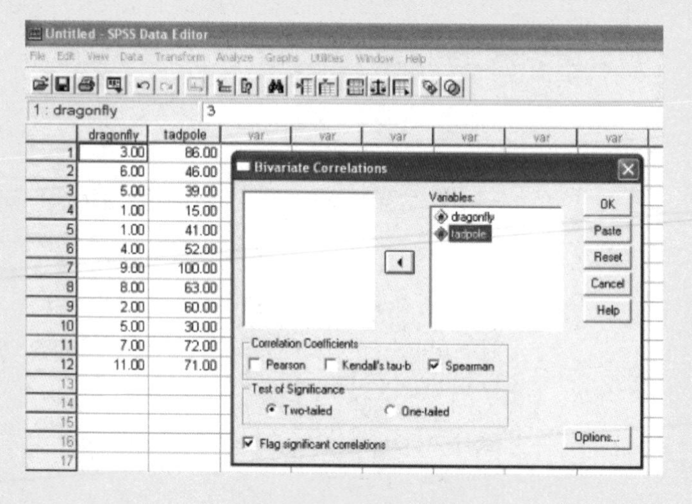

Finally click onto **OK** to run the test. SPSS will come up with the following results.

Correlations

			Dragonfly	Tadpole
Spearman's rho	Dragonfly	Correlation coefficient	1.000	0.547 ←
		Sig. (2-tailed)	.	0.065 ←
		N	12	12
	Tadpole	Correlation coefficient	0.547	1.000
		Sig. (2-tailed)	0.065	.
		N	12	12

SPSS gives the Spearman rank correlation, ρ, here 0.547.

Using MINITAB

MINITAB does not calculate the Spearman correlation directly. Instead, you must convert your data into ranks, and then perform a conventional Pearson correlation analysis (see Section 6.7) on the ranked data. To convert the two columns of data to ranks. Click on the **Calc** menu, and click onto the **Calculator** bar. This opens the **Calculator** dialogue box. For each column put the expression **RANK('Dragonfly')** and **RANK('Tadpole')** into the **Expression** box, and put C3 or C4 into the **Store result in variable** box. The final completed data and box are shown below.

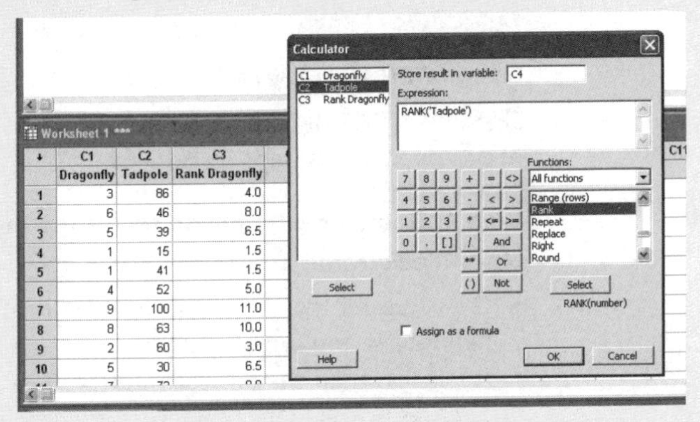

Click on **OK** to complete the ranking of the data. Then move onto the **Basic Statistics** bar and click on **Correlation**. MINITAB will produce the **Correlation** dialogue box. Move **Rank Dragonfly** and **Rank Tadpole** into the **Variables** box. The completed box and data are shown below.

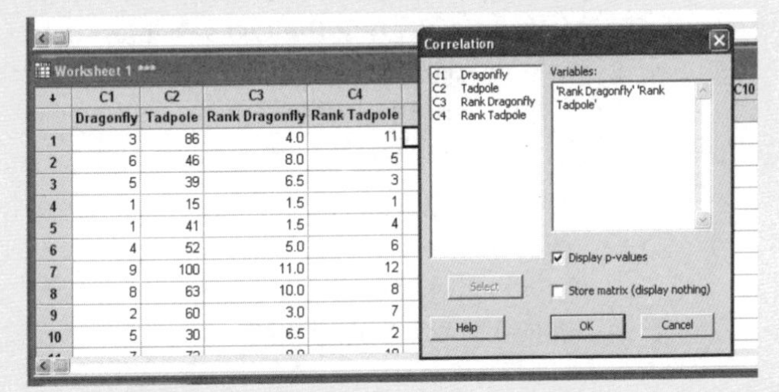

Finally click onto **OK** to run the test. MINITAB will come up with the following results.

Correlations: Rank Dragonfly, Rank Tadpole

```
Pearson correlation of Rank Dragonfly and Rank Tadpole = 0.547 ←

P-Value = 0.065
            ↑
```

The value given (here 0.547) is the Spearman rank correlation, ρ.

Step 3: Calculating the significance probability

You must calculate the probability P that the absolute value of the test statistic ρ would be **greater than or equal to** ρ if the null hypothesis were true.

Using a calculator

You must compare your value of ρ with the critical value of the Spearman correlation coefficient for $N - 2$ degrees of freedom, where N is the number of pairs of observations. This is given in Table S7 at the end of the book.

Here looking up ρ for $(12 - 2) = 10$ degrees of freedom gives a critical value of $\rho = 0.648$.

Using SPSS and MINITAB

SPSS and MINITAB both show that the probability Sig. (two-tailed) or P-value of getting such test statistics is 0.065.

Step 4: Deciding whether to reject the null hypothesis

Using a calculator

- If ρ is greater than or equal to the critical value, you must reject the null hypothesis. You can say that the two variables show significant correlation.
- If ρ is less than the critical value, you cannot reject the null hypothesis. There is no evidence of a rank correlation between the two variables.

Here $\rho = 0.547 < 0.648$.

Using SPSS and MINITAB

- If Sig. (two-tailed) or P-value ≤ 0.05 you must reject the null hypothesis. You can say that the two variables show significant correlation.
- If Sig. (two-tailed) or P-value > 0.05 you have no evidence to reject the null hypothesis. There is no evidence of a rank correlation between the two variables.

Here Sig. (two-tailed) = P-value = $0.065 > 0.050$.

Therefore we have no evidence to reject the null hypothesis. Though it looks like there is a positive association between tadpole and dragonfly density it is not significant.

6.9.4 Presenting the results

These results are best presented by producing a scatterplot. For example the results of tadpole and dragonfly density can be presented as in Figure 6.10.

The graph should be referred to in the text of the results section where the results of the Spearman correlation analysis should be given. Here you would say something like

> The relationship between the density of dragonfly and tadpole density is given in Figure 6.10. A Spearman correlation showed that there was no significant relationship between them ($\rho_{10} = 0.547$ $p = 0.065$).

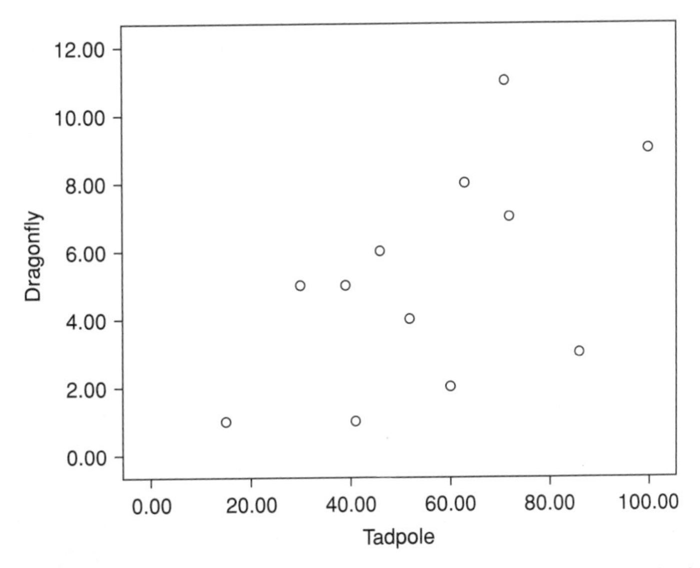

Figure 6.10 Graph showing the relationship between the density of tadpoles and dragonfly larvae in 12 ponds.

6.10 Self-assessment problems

Problem 6.1

Which way round would you plot the following data?

(a) Cell number of an embryo and time since fertilisation
(b) Pecking order of hens and of their chicks
(c) Height and body weight of women
(d) Length and breadth of limpets

Problem 6.2

A study of the density of stomata in vine leaves of different areas came up with the following results. Calculate the correlation coefficient r between these two variables and determine whether this is a significant correlation. What can you say about the relationship between leaf area and stomatal density?

Leaf area (mm^2)	45	56	69	32	18	38	48	26	60	51
Stomatal density (mm^{-2})	36	28	27	39	56	37	32	45	24	31

Problem 6.3

In a survey to investigate why bones become more brittle in older women, the density of bone material was measured in 24 post-menopausal women of contrasting ages. Bone density is given as a percentage of the average density in young women.

Age (years)	43	49	56	58	61	63	64	66	68	70	72	73
Relative bone density	108	85	92	90	84	83	73	79	80	76	69	71
Age (years)	74	74	76	78	80	83	85	87	89	92	95	98
Relative bone density	65	64	67	58	50	61	59	53	43	52	49	42

(a) Plot the data.

(b) Carry out a regression analysis to determine the relationship between age and bone density. Does bone density change significantly with age?

(c) Calculate the expected bone density of women of age 70.

Problem 6.4

In an experiment to examine the ability of the polychaete worm *Nereis diversicolor* to withstand zinc pollution, worms were grown in solutions containing different concentrations of zinc and their internal zinc concentration was measured. The following results were obtained.

$\log_{10} [Zn]_{water}$	1.96	2.27	2.46	2.65	2.86	2.92	3.01	3.24	3.37	3.49
$\log_{10} [Zn]_{worm}$	2.18	2.23	2.22	2.27	2.25	2.30	2.31	2.34	2.36	2.35

(a) Plot the data.

(b) Carry out a regression analysis to determine how zinc in the solution affects the concentration within the worm. If *Nereis* did not actively control its level of zinc, the concentrations inside and outside would be equal and the slope of the regression line would be 1. Work out the t value which compares a slope of 1 with the slope of the line you obtained, hence determine whether *Nereis* does actively control its zinc level.

Problem 6.5

A study of the effect of seeding rate on the yield of wheat gave the following results.

Seeding rate (m^{-2})	50	80	100	150	200	300	400	500	600	800
Yield (tonnes)	2.5	3.9	4.7	5.3	5.6	5.9	5.4	5.2	4.6	3.2

(a) Plot a graph of yield against seeding rate.

(b) Carry out regression analysis to determine whether there is a significant linear relationship between seeding rate and yield.

(c) What can you say about the relationship between seeding rate and yield?

Problem 6.6

(a) The logarithms of the wing area A of birds and their body length L are found to be related by the straight line relationship $\log_{10} A = 0.3 + 2.36 \log_{10} L$. What is the relationship between A and L?

(b) The natural logarithm of the numbers of cells N in a bacterial colony is related to time T by the equation $\log_e N = 2.3 + 0.1T$. What is the relationship between N and T?

Problem 6.7

An investigation was carried out into the temperature dependence of the metabolism of a species of coecilian (a worm-like amphibian). A captive animal was kept at temperatures ranging from 0 to 30° C at intervals 2° C and its metabolic rate determined by measuring the rate of output of carbon dioxide. The following results were obtained.

Temperature	0	2	4	6	8	10	12	14
CO_2 production (ml/min)	0.35	0.43	0.45	0.55	0.60	0.78	0.82	0.99
Temperature	16	18	20	22	24	26	28	30
CO_2 production (ml/min)	1.32	1.43	1.64	1.71	2.02	2.35	2.99	3.22

Transform the data by taking natural logarithms of CO_2 production and use regression analysis to examine the relationship between temperature and metabolic rate.

Problem 6.8

It was thought that the dominance of male rats might be related to the levels of testosterone in their blood. Therefore encounters between a total of 20 rats were observed and the dominance order, from 1 for the top rat to 20 for the bottom was then worked out. Blood samples were also taken from each rat to measure its testosterone level. The following results were obtained.

Dominance	1	2	3	4	5	6	7	8	9	10
Testosterone	7.8	6.7	7.3	6.8	6.2	8.1	7.8	6.5	6.9	7.0

Dominance	11	12	13	14	15	16	17	18	19	20
Testosterone	6.7	6.4	6.3	5.8	7.6	6.7	6.6	7.1	6.4	6.5

Carry out a Spearman rank correlation analysis to determine whether there was a significant association between testosterone and dominance.

7 Dealing with categorical data

7.1 Introduction

Often in biology you do not take **measurements** on organisms or other items, but classify them into different **categories**. For instance, birds belong to different species and have different colours; habitats (and Petri dishes) can have particular species present or absent; and people can be healthy or diseased. You cannot sensibly assign numbers to such arbitrarily defined classes; green is not larger in any real sense than yellow! For this reason you cannot use any of the statistical tests we examined in Chapters 4, 5 or 6 that look for differences or relationships between measurements.

Instead, this categorical data is best quantified by counting the numbers of observations in the different categories. This will allow you to estimate the **frequency** with which each character state turns up. This data can then be used to answer one of three questions:

frequency
The number of times a particular character state turns up.

- We might want to know whether the character frequencies in a single group are different from expected values. Do rats in a maze turn to the right rather than left at a different frequency from the expected 1:1? Or is the frequency of rickets different in a small mining town from that in the general population?
- We might want to know whether the character frequencies in two or more groups are different from each other. In other words, are certain characteristics associated with each other? For example, is smoking more common in men than women? Or do different insect species preferentially visit different species of flower?
- We might want to know if certain characteristics are associated with other measurements or ranks. For example is smoking more common in taller people? Or are lower-ranked birds more likely to be eaten by predators?

7.2 The problem of variation

At first glance it might seem easy to tell whether character frequencies are different. When looking at a sample of sheep, if we found that eight were black and six white, we might conclude that black ones were commoner than white. Unfortunately, there might easily have been the same number of black and white sheep in the population and we might just have picked more black ones by chance.

binomial distribution
The pattern by which the sample frequencies in two groups tends to vary.

mean (μ)
The average of a population. The estimate of μ is called $\bar{x}$.

standard deviation
(σ)A measure of spread of a group of measurements: the amount by which on average they differ from the mean. The estimate of σ is called s.

A character state is, in fact, unlikely to appear at exactly the same frequency in a small sample as in the whole population. Let's examine what happens when we take samples of a population of animals, 50% of which are white and 50% black. In a sample of 2 there is only a 50% chance of getting a 1:1 ratio; the other times both animals would be either black or white. With 4 animals there will be a 1:1 ratio only 6 times out of 16; there will be a 3:1 or 1:3 ratio 4 times out of 16 and a 4:0 or 0:4 ratio once every 16 times.

As the number of animals in the sample increases, the most likely frequencies are those closer and closer to 1:1, but the frequency will hardly ever equal 1:1 exactly. In fact the probability distribution will follow an increasingly tight **binomial distribution** (Figure 7.1) with **mean** $\bar{x}$ equal to $n/2$, where n is the sample size and **standard deviation** s approaching $\sqrt{(n/2)}$. The probability that the ratio is near 1:1 increases, and the chances of it being further away decreases. However, there is always a finite, if increasingly tiny, chance of getting all white animals.

Things get more complex if the expected frequencies are different from 1:1 and if there are a larger number of categories, but essentially the same pattern will occur: as the sample size increases, the frequencies will tend to approach, but seldom equal, the frequencies in the population. The probability of obtaining frequencies similar to that of the population rises, but there is still a finite probability of the frequencies being very different. As expected, in a population where a character occurs at a frequency p, the mean frequency at which it will turn up in a sample is also p. However, the standard deviation, s (also confusingly called the standard error) is given by the rather complex formula

$$s = \sqrt{p(1 - p)/(n - 1)} \tag{7.1}$$

So if the results from a sample are different from the expected frequency, you cannot be sure this is because the population you are sampling is really different. Even if you sampled 100 animals and all were white, the population still might have contained a ratio of 1:1; you might just have been very unlucky. However, the greater the difference and the larger the sample, the less likely this becomes. To determine whether differences from an expected frequency

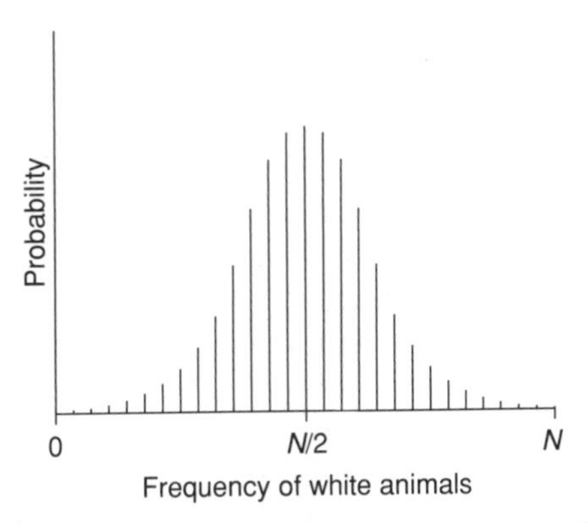

Figure 7.1 The binomial distribution. Probabilities of choosing different numbers of white animals in a sample of size N from a population with 50% white animals.

chi-squared (χ2)
A statistical test which determines whether there are differences between real and expected frequencies in one set of categories, or associations between two sets of categories.

are likely to be real, there are several types of test you can use: the **chi-squared (χ^2) test**, the **G test**, the **Kolgomorov–Smirnov one-sample test** and **Fisher's exact test**. However, by far the most commonly used is the **chi-squared (χ^2) test**, which we will examine here. There are two main types of χ^2 test:

- The χ^2 test for differences
- The χ^2 test for association

7.3 The χ^2 test for differences

7.3.1 Purpose

To test whether character frequencies are different from expected values. It is best used when you expect numbers in different categories to be in particular ratios. Here are some examples:

- Experiments in Mendelian genetics
- Maze or choice chamber experiments
- Examining sex ratios
- Comparing your results with figures from the literature
- Comparing figures from a small sample with national averages

7.3.2 Rationale

The test calculates the chi-squared statistic (χ^2); this is a measure of the difference between the observed frequencies and the expected frequencies. Basically, the larger χ^2, the less likely the results could have been obtained by chance if the population frequency was the expected one.

The χ^2 statistic is given by the simple expression

$$x^2 = \sum \frac{(O - E)^2}{E} \tag{7.2}$$

where O is the observed frequency and E is the expected frequency for each character state. The larger the difference between the frequencies, the larger the value of χ^2 and the less likely it is that observed and expected frequencies are different just by chance. Similarly, the bigger the sample, the larger O and E, hence the larger the value of χ^2; this is because of the squared term in the top half of the fraction. So the bigger the sample you take, the more likely you will be to detect any differences.

The greater the number of possible categories there are, the greater the number of degrees of freedom; this also tends to increase χ^2. The distribution of χ^2 has been worked out for a range of degrees of freedom, and Table S3 (at the end of the book) gives the critical values of χ^2 above which there is less than a 5%, 1% or 0.1% probability of getting the observed values by chance.

7.3.3 Carrying out the test

To see how to carry out a χ^2 test for differences it is best to run through a straightforward example.

Example 7.1

In a Mendelian genetics experiment, F1 hybrids of smooth and wrinkled peas were crossed together. The following results were obtained:

Number of smooth peas = 69
Number of wrinkled peas = 31

Test whether the ratio of smooth to wrinkled peas in the 100 progeny is different from the 3:1 ratio predicted by Mendelian genetics. Carrying out the test involves the usual four stages.

Step 1: Formulating the null hypothesis

The null hypothesis is that the ratio obtained was equal to that expected. Here the null hypothesis is that the ratio of smooth:wrinkled peas equals 3:1.

Step 2: Calculating the test statistic

Using a calculator

The first thing to do is to calculate the expected values. Here three-quarters should be smooth and one quarter wrinkled. Since there are 100 progeny,

$$\text{Expected number of smooth} = (3 \times 100)/4 = 75$$
$$\text{Expected number of wrinkled} = (1 \times 100)/4 = 25$$

So we have

	Observed	Expected	Observed–expected
Smooth	69	75	–6
Wrinkled	31	25	6

Now we can caclulate χ^2.

$$\chi^2 = \sum \frac{(O - E)^2}{E}$$
$$\chi^2 = \frac{(69 - 75)^2}{75} + \frac{(31 - 25)^2}{25}$$
$$\chi^2 = 36/75 + 36/25 = 0.48 + 1.44 = 1.92$$

Using SPSS

Conventionally, carrying out χ^2 tests for differences in SPSS involves putting in data about each organism or item and then carrying out the test. Since sample sizes are invariably large, this approach could be extremely time-consuming. Fortunately, there is a quicker approach which involves weighting the data you enter.

As usual the first thing is to enter your data correctly. In this case you will have to enter data into two columns. In the first column you should enter subscripts for each of the **categories** (here 1 and 2). You can label these categories smooth and wrinkled. In the second column you should put the numbers in each of the categories. You should call this **weighting.**

To weight the cases, go into **Data** and click on **Weight Cases**. This will bring up the **Weight Cases** dialogue box. Click on **Weight Cases** by and enter **Weighting** into the **Frequency Variable** box. The completed dialogue box and data are shown below.

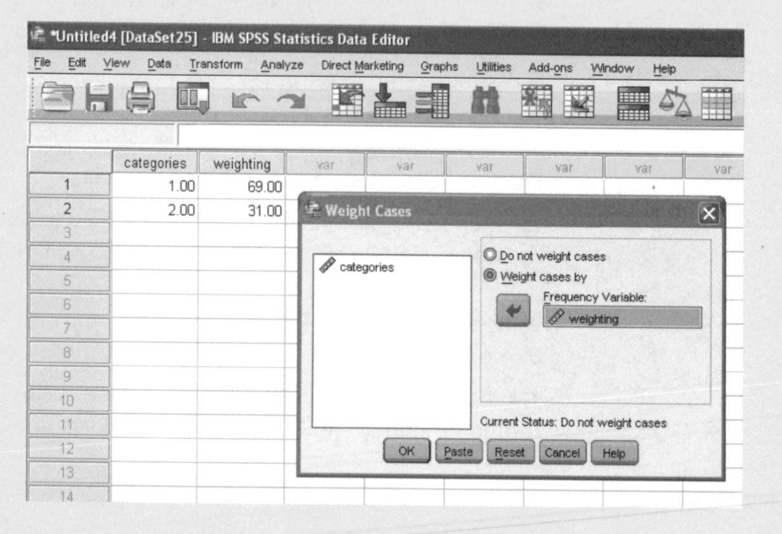

Click on **OK** to weight the data. Now you can carry out the test (which you could also do if you had a column with 69 1's and 31 s's!). Click on **Analyse**, go into **Nonparametric Tests** onto **Legacy Dialogue** and click

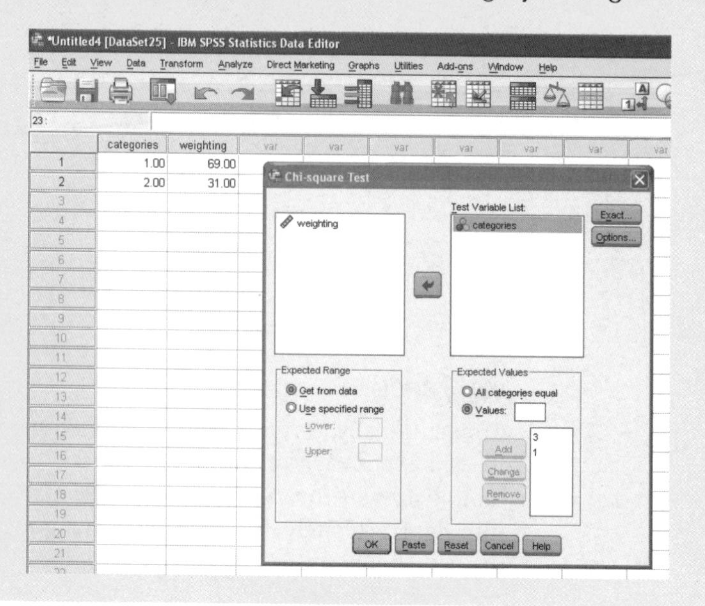

on **Chi-Square.** This will bring up the **Chi-Square Test** dialogue box. Put the **Categories** column into the **Test Variable List** box. Finally you need to enter the expected ratios of your two categories (here 3:1). To do this click on **Values** in the **Expected Values** box and type in 3 (the expected value for category 1) and click on **Add** to enter it. The enter 1 and **Add** to do the same for category 2. The completed data and dialogue box are shown in previous page bottom.

Finally click on **OK** to run the test. SPSS will print the following tables:

Categories

	Observed *N*	Expected *N*	Residual
1.00	69	75.0	−6.0
2.00	31	25.0	6.0
Total	100		

Test statistics

	Categories
Chi-square[a]	1.920 ←
df	1
Asymp. Sig.	0.166 ←

[a]0 cells (0.0%) have expected frequencies less than 5. The minimum expected cell frequency is 25.0.

Observed and expected values are given in the first table, the value of χ^2 (here 1.920) in the second.

Using MINITAB

It is straightforward to carry out χ^2 tests for differences in MINITAB. Put the frequencies into the first column and the expected *proportions* (here 0.75 and 0.25) into the second column. Call them something like **observed** and **expected.** Next click on the **Stat** menu, move onto the **Tables** bar and

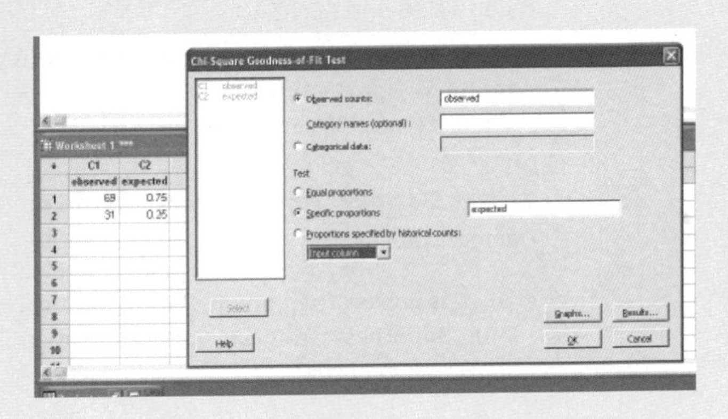

click on **Chi-Square Goodness-of-Fit Test (One Variable)**. MINITAB will produce the **Chi-Square Goodness-of-Fit Test** dialogue box. Move **observed** into the **observed counts** box. The click on **Specific proportions** and move **expected** into the box. The completed data and box are shown in previous page bottom.

Finally click on **OK**. MINITAB will come up with the following results

Chi-Square Goodness-of-Fit Test for Observed counts in variable: observed

Category	Observed	Test Proportion	Expected	Contribution to Chi-Sq
1	69	0.75	75	0.48
2	31	0.25	25	1.44

N	DF	Chi-Sq	P-Value
100	1	1.92	0.166

Observed and expected values are given in the first table, the value of χ^2 (here 1.92) in the second. MINITAB will also give you some not very useful tables. You may stop it doing this by going into the **Graphs** dialogue box and unticking the two bar charts.

Step 3: Calculating the significance probability

Using a calculator

You must compare your value of χ^2 with the critical value of the χ^2 statistic for $(N - 1)$ degrees of freedom, where N is the number of character states, and at the 5% level.

Here for $2 - 1 = 1$ degree of freedom, the critical value of $\chi^2 = 3.84$.

Using SPSS and MINITAB

SPSS and MINITAB directly give the significance probability, here called **Asymp. Sig.** or ***P*-value**

Here Asymp. Sig. = P-value = 0.166

Step 4: Deciding whether to reject the null hypothesis

Using a calculator

- If χ^2 is greater than or equal to the critical value, you must reject the null hypothesis. You can say that the distribution is significantly different from expected.

- If χ^2 is less than the critical value, you cannot reject the null hypothesis. You have found no significant difference from the expected distribution.

Here, $\chi^2 = 1.92 < 3.84$.

Using SPSS and MINITAB

- If Asymp. Sig. or *P*-value $\leq$ 0.05 you must reject the null hypothesis. You can say that the distribution is significantly different from expected.
- If Asymp. Sig. or *P*-value > 0.0 5 you have no evidence to reject the null hypothesis. You have found no significant difference from the expected distribution.

Here Asymp. Sig. = *P*-value = 0.166 > 0.05.

Therefore we have no evidence to reject the null hypothesis. The relative frequencies of smooth and wrinkled peas are not significantly different from the expected 3:1 ratio.

7.3.4 Presenting the results

You can usually present the results of χ^2 tests for differences in the text. For instance the results of the current example might be presented thus.

> Of the 100 progeny, 69 produced smooth and 31 produced wrinkled peas, which was not significantly different from the expected 3:1 ratio ($\chi_1{}^2 = 1.92$, $P > 0.05$).

7.4 The χ^2 test for association

7.4.1 Purpose

To test whether the character frequencies of two or more groups are different from each other. In other words, to test whether character states are associated in some way. It is used when there is no expected frequency. Here are some examples:

- Ecological surveys: Are different species found in different habitats?
- Medical surveys: Are infection rates different for people in different blood groups?
- Sociological surveys: Do men and women have a different probability of smoking?

7.4.2 Rationale

The test investigates whether the distribution is different from what it would be if the character states were distributed randomly among the population. The tricky part is determining the expected frequencies. Once these have been determined, however, the value of χ^2 is found in just the same way as for the χ^2 test for differences, using equation 6.2.

7.4.3 Carrying out the test

To see how to carry out a test, it is best as usual to run through an example.

Example 7.2

A sociological study found that out of 30 men, 18 were smokers and 12 non-smokers, while of the 60 women surveyed, 12 were smokers and 48 were non-smokers. Test whether the rates of smoking are significantly different between the sexes.

Solution

Step 1: Formulating the null hypothesis

The null hypothesis is that there is no association between the character states. Here, the null hypothesis is that there is no association between gender and the likelihood that a person will smoke, so men and women are equally likely to smoke.

Step 2: Calculating the test statistic

Using a calculator

This is a long-winded process because before we can calculate χ^2 we must first calculate the expected values for each character state if there had been no association. The first stage is to arrange the data in a **contingency table**.

contingency table
A table showing the frequencies of two sets of character states, which allows you to calculate expected values in a χ^2 test for association.

	Smoking	Non-smoking	Total
Men	18	12	30
Women	12	48	60
Total	30	60	90

It is now possible to calculate the frequencies we would expect *if there had been no association between smoking and gender.* Of the total number of people examined, one-third (30) were men, and one-third (30) of all people smoked. Therefore if the same proportion of men smoked as in the general population, you would expect one-third of all men (10) to be smokers. Hence 20 men should be non-smokers. Similarly, of the 60 women only one-third (20) should be smokers and 40 should be non-smokers.

A general expression for the expected number E in each cell of the contingency table is given by

$$E = \frac{\text{Column total} \times \text{Row total}}{\text{Grand total}} \tag{7.3}$$

where the grand total is the total number of observations (here 90). Therefore, the expected value for male smokers is found by multiplying its row total (30) by the column total (30) and dividing by 90, to give 10. These expected values are then put into the contingency table, written in parentheses. It is now straightforward to calculate χ^2 using equation 7.2.

	Smoking	Non-smoking	Total
Men	18 (10)	12 (20)	30
Women	12 (20)	48 (40)	60
Total	30	60	90

$$\chi^2 = \sum \frac{(O - E)^2}{E}$$

$$\chi^2 = \frac{(18 - 10)^2}{10} + \frac{(12 - 20)^2}{20} + \frac{(12 - 20)^2}{20} + \frac{(48 - 40)^2}{40}$$

$$\chi^2 = 6.4 + 3.2 + 3.2 + 1.6 = 14.4$$

Using SPSS

Once again SPSS can calculate χ^2 if you enter each person separately in a large data sheet, with separate columns for each characteristic. Since there are 90 people here this involves entering 180 numbers. However, you can also do it much quicker by weighting the data as in Example 7.1.

First, you will need to put the data into **two columns,** one for **gender** (give men the subscript 1 and the women 2), the other for **smoking** (give non-smokers the subscript 0 and smokers 1). Doing it longhand, for the **gender** column give the first 30 people the subscript 1 (meaning men), and for 31–90 give them subscript 2 (meaning women). In the **smoking** column give the first 18 men the subscript 1 (meaning smoking) and the final 12 the subscript 0 (meaning non-smoking). Finally give the first 12 women the subscript 1 and the final 48 women the subscript 0.

To do it quicker produce three columns: a **gender** column, a **smoking** column and a **weighting** column. You need just four rows, one for each combination of categories, and the weighting column should give the numbers in each of the categories. The completed columns are shown below.

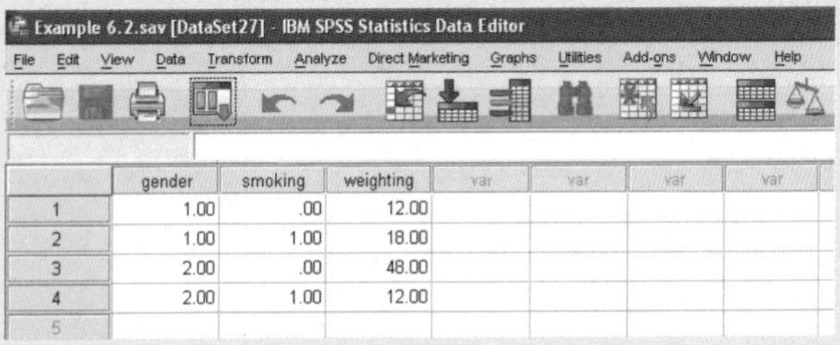

Example 6.2.sav [DataSet27] - IBM SPSS Statistics Data Editor

File Edit View Data Transform Analyze Direct Marketing Graphs Utilities Add-ons Window Help

	gender	smoking	weighting	var	var	var	var
1	1.00	.00	12.00				
2	1.00	1.00	18.00				
3	2.00	.00	48.00				
4	2.00	1.00	12.00				
5							

To weight the data click on **Data** and then on **Weight Cases**. You then weight the data by clicking on **Weight Cases** by and moving **Weighting** into the **Frequency Variable** box. The completed dialogue box is shown below.

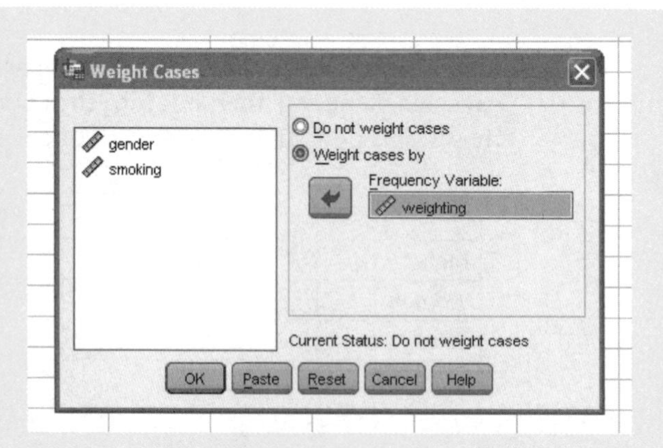

Finally click on **OK** to weight the cases.

To carry out the test, click on **Analyse,** move onto **Descriptive Statistics** and click on **Crosstabs**. This will bring up the **Crosstabs** dialogue box. Put **gender** into the **Row(s)** box and **smoking** into the **Column(s)** box. Next, click on the **Statistics** box to bring up the **Crosstab: Statistics** dialogue box and tick **Chi-square.** Your dialogue boxes and data screen will look like the following.

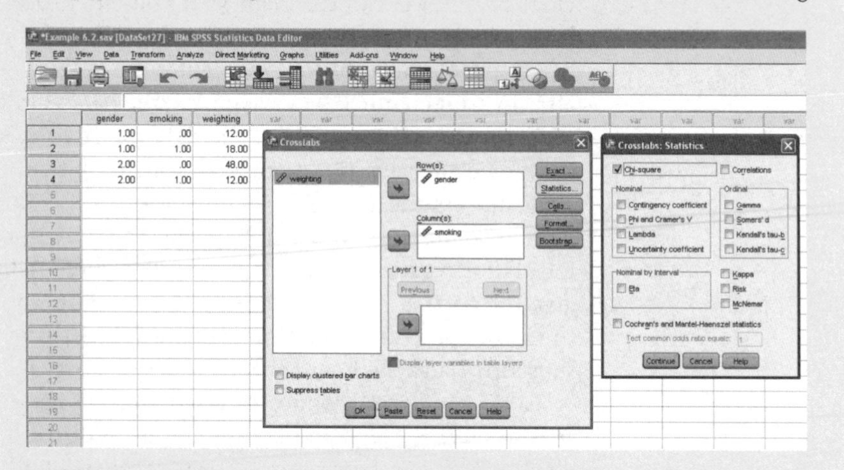

Click on **Continue** then click on **Cells** to bring up the **Crosstabs: Cell Display** dialogue box. Make sure both **Observed** and **Expected** are ticked as below.

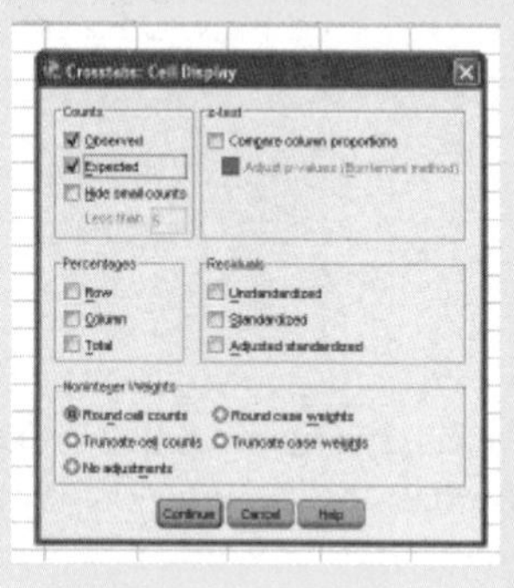

Finally click on **Continue** and **OK** to run the test. SPSS will come up with the following useful output.

Crosstabs

Gender * smoking crosstabulation

			Smoking		Total
			0.00	1.00	
Gender	1.00	Count	12	18	30
		Expected count	20.0	10.0	30.0
	2.00	Count	48	12	60
		Expected count	40.0	20.0	60.0
Total		Count	60	30	90
		Expected count	60.0	30.0	90.0

Chi-square tests

	Value	df	Asymp. Sig. (2-sided)	Exact Sig. (2-sided)	Exact Sig. (1-sided)
Pearson chi-square	14.400[b]	1	0.000		
Continuity correction[a]	12.656	1	0.000		
Likelihood ration	14.144	1	0.000		
Fisher's exact Test				0.000	0.000
Linear-by-linear Association	14.240	1	0.000		
N of valid cases	90				

[a]Computed only for a 2×2 table
[b]0 cells (0.0%) have expected count less than 5. The minimum expected count is 10.00.

The first table is the contingency table, and the statistic you require is the **Pearson chi-square**. Here Pearson chi-square = 14.400.

Using MINITAB

It is straightforward to carry out χ^2 tests for associations in MINITAB. Put the frequencies into the data sheet in the same arrangement as in the con-

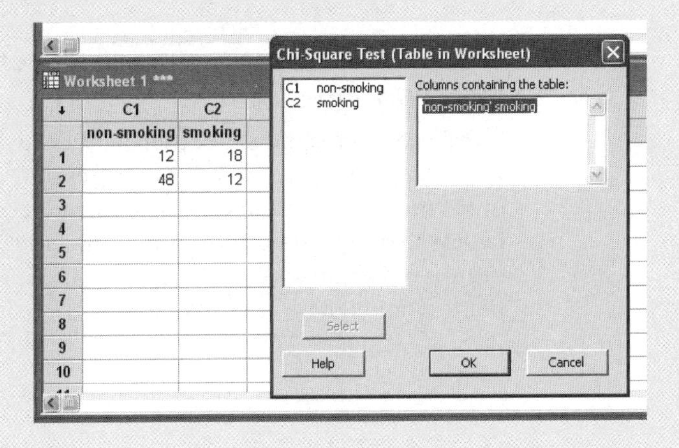

tingency table. Call the first column **non-smoking** and the second **smoking**. Next click on the **Stat** menu, move onto the **Tables** bar and click on **Chi-Square Test (Two-Way Table in Worksheet).** MINITAB will produce the **Chi-Square Test (Table in Worksheet)** dialogue box. Move the two columns into the **Columns containing the table:** box. The completed data and box are shown in previous page bottom.

Finally click on **OK** to run the test. MINITAB will come up with the following output.

Chi-Square Test: non-smoking, smoking

```
Expected counts are printed below observed counts

Chi-Square contributions are printed below expected counts

        non-smoking  smoking  Total

    1            12       18     30

              20.00    10.00

              3.200    6.400

    2            48       12     60

              40.00    20.00

              1.600    3.200

Total            60       30     90

Chi-Sq = 14.400, DF = 1, P-Value = 0.000
```

The first table is the contingency table with the expected values and values of χ^2 for each cell. The line below gives the total χ^2 value (χ^2) which is 14.400.

Step 3: Calculating the significance probability

Using a calculator

To calculate the significance probability you must look up the value that χ^2 must exceed at $(R - 1) \times (C - 1)$ degrees of freedom, where R is the number of rows in the table and C is the number of columns, for the probability to be less than 5%. Here there are 2 rows and 2 columns, so the number of degrees of freedom $= (2 - 1) \times (2 - 1) = 1$. The critical value of χ^2 for 1 degree of freedom = 3.84.

Using SPSS and MINITAB

SPSS gives you the significance probability (here called **Asymp. Sig. (2-sided)**) or ***P*-value** directly. Here Asymp. Sig. (2-sided) = P-value = 0.000.

Step 4: Deciding whether to reject the null hypothesis

Using a calculator

- If χ^2 is greater than or equal to the critical value, you must reject the null hypothesis. You can say that the distribution is significantly different from expected, hence there is a significant association between the characters.
- If χ^2 is less than the critical value, you cannot reject the null hypothesis. You have found no significant difference from the expected distribution, hence no evidence of an association between the characters.

Here $\chi^2 = 14.4 > 3.84$.

Using SPSS and MINITAB

- If Asymp. Sig. (2-sided) or P-value ≤ 0.05 you should reject the null hypothesis. Therefore you can say that the slope (or intercept) is significantly different from zero.
- If Asymp. Sig. (2-sided) or P-value > 0.05 you have no evidence to reject the null hypothesis. Therefore you can say that the slope (or intercept) is not significantly different from zero.

Here Asymp. Sig. (2-sided) = P-value = $0.000 < 0.05$.

Therefore you can reject the null hypothesis. You can say there is a significant association between sex and smoking. In other words, the two sexes are different in the frequency with which they smoke. In fact men smoke more than women.

You can tell even more about your results by looking at the χ^2 values for each of the cells. The larger the value, the more the results for the cell differ from the expected results. In this example χ^2 for male smokers is by far the largest at 6.4. Therefore we can say that in particular more men smoke than one would expect.

7.4.4 Presenting the results

When presenting the results of χ^2 tests for association, you may be able to present the results in the text, particularly if it is a 2×2 table as in the example we have just seen. You might write

> A significantly greater proportion of men smoked than women ($\chi_1{}^2 = 14.4$, $P < 0.001$); of the 30 men questioned, 18 were smokers and 12 non-smokers, whereas of the 60 women only 12 smoked, with 48 non-smokers.

Table 7.1 The numbers of men and women and their smoking status. The table gives both observed and expected (in brackets) numbers

Gender	Smoking status	Non-smoking	Smoking total
Men	12 (20)	18 (10)	30
Women	48 (40)	12 (20)	60
Total	60	30	90

However, particularly when there are more categories it is usually best to present the completed contingency table so that readers can see what has happened. For instance the results of Example 7.2 can be presented as in Table 7.1.

It could be referred to in the text as follows.

> Table 7.1 shows the numbers of men and women who were smokers or non-smokers. A χ^2 test showed that there were significant differences in the incidence of smoking in the two genders (${\chi_1}^2 = 14.4$, $P < 0.001$). Men were more likely to smoke than women.

7.4.5 The Yates continuity correction

Note that for 2×2 contingency tables like this, some people believe that you should make a correction for the fact that you can only get integer values in each category. This would tend to result in an overestimation of χ^2. To correct for this one can add or subtract 0.5 from each observed value to make them closer to the expected values. This **continuity correction** results in a lower value for χ^2, and can be read off the SPSS output. In our example $\chi^2 = 12.656$ rather than 14.400. Opinions are divided about whether this is a valid procedure. In any case, this correction can only be made in 2×2 tables.

7.5 Validity of χ^2 tests

1. You must only carry out χ^2 tests on the raw numbers of observations that you have made. *Never* use percentages. This is because the larger the number of observations, the more likely you are to be able to detect differences or associations with the χ^2 test.
2. Another point about sample size is that χ^2 tests are only valid if all expected values are larger than 5. If any expected values are lower than 5, there are two possibilities:

   - You could combine data from two or more groups, but only if this makes biological sense. For instance, different species of fly could be combined in Problem 7.4 because flies have more in common with each other than with the other insects studied.
   - If there is no sensible reason for combining data, small groups should be left out of the analysis.

Example 7.3

In a survey of the prevalence of a heart condition in 300 people of different races, the following results were obtained.

	White	Asian	Afro Carribean	Mixed race
Have the condition	35	9	13	3
Healthy	162	43	29	6

Do people from different races differ in their likelihood of having the condition? In other words, is there an association between race and the condition?

Solution

The first thing to notice is that there are too few mixed race people in the survey. The expected value for mixed race people with the condition, for instance is $(9 \times 60)/300 = 1.8$, which is well below 5. There is no justification for putting mixed race people into any other **category** so they must be ignored. Now we can carry out a χ^2 test for association for the other three races.

category
A character state which cannot meaningfully be represented by a number.

Step 1: Formulating the null hypothesis

The null hypothesis is that there is no association between race and the condition.

Step 2: Calculating the test statistic

The contingency table must be reduced and the expected values calculated.

	White	**Asian**	**Afro Carribean**	**Total**
Have the condition	35 (38.6)	9 (10.2)	13 (8.2)	57
Healthy	162 (158.4)	43 (41.8)	29 (33.8)	234
Total	197	52	42	291

Now χ^2 can be found using equation 6.2,

$$\chi^2 = \frac{(35 - 38.6)^2}{38.6} + \frac{(9 - 10.2)^2}{10.2} + \frac{(13 - 8.2)^2}{8.2} + \frac{(162 - 158.4)^2}{158.4} + \frac{(43 - 41.8)^2}{41.8} + \frac{(29 - 33.8)^2}{33.8}$$

$$= 0.34 - 0.14 + 2.81 + 0.08 + 0.03 + 0.68$$

$$= 4.08.$$

Step 3: Calculating the significance probability

You must look up the value that χ^2 must exceed at $(3 - 1) \times (2 - 1) = 2$ degrees of freedom. The critical value of χ^2 for 2 degrees of freedom = 5.99.

Step 4: Deciding whether to reject the null hypothesis

Here $\chi^2 = 4.08 < 5.99$ so you have no evidence to reject the null hypothesis. You can say there is no significant difference in the incidence of the condition between people of the different races.

7.6 Logistic regression

7.6.1 Purpose

logistic regression
A statistical test which analyses how a binary outcome is affected by other numerical characteristics.

The purpose of **logistic regression** is to test whether the frequencies of particular traits (usually binary traits such as diseased or fit, eaten or ignored, yes or no) are influenced by other traits of the same organisms or cells. Logistic regression essentially does the same thing as the χ^2 test for association, but it can also investigate how these traits are influenced by ranked and measurement data, as well as categorical data. Here are some examples:

- Ecological surveys: Are the rates of predation different on animals with different levels of protective chemicals?
- Medical surveys: Is the rate of type 2 diabetes different for people with different body mass indices?
- Sociological surveys: Do people with different incomes have a different probability of smoking?

7.6.2 Rationale

The test investigates whether the distribution is different from what it would be if the character states were distributed randomly among the population. It looks at how much the distribution changes with the other trait, and performs a *t* test just like that done in linear regression to see whether the trait has a significant effect.

7.6.3 Carrying out the test

To see how to carry out a test, it is best as usual to run through an example.

Example 7.4

In a study into the effectiveness of camouflage on the predation of model caterpillars, 50 unrealistic models, 50 semi-camouflaged and 50 completely camouflaged models were put into a cage into which starlings were introduced. After one hour, 27 unrealistic, 19 semi-camouflaged and 17 completely camouflaged models had been eaten. The results can be described in the following contingency table.

	Unrealistic	Semi-camouflaged	Completely camouflaged	Total
Eaten	27	19	17	62
Uneaten	23	31	33	88
Total	50	50	50	150

Test whether camouflage significantly affects predation.

Solution

Step 1: Formulating the null hypothesis

The null hypothesis is that there is no association between the character state and the second trait. Here, the null hypothesis is that the level of camouflage has no effect on predation, so the different models are equally likely to be eaten.

Step 2: Calculating the test statistic

Using a calculator

You cannot easily perform this test on a calculator.

Using SPSS

The usual way (and if you are using measurements the only way) of putting data into SPSS is to enter each model separately in a large data sheet, with separate columns for the two characteristics: its realism (1, 2 or 3), and

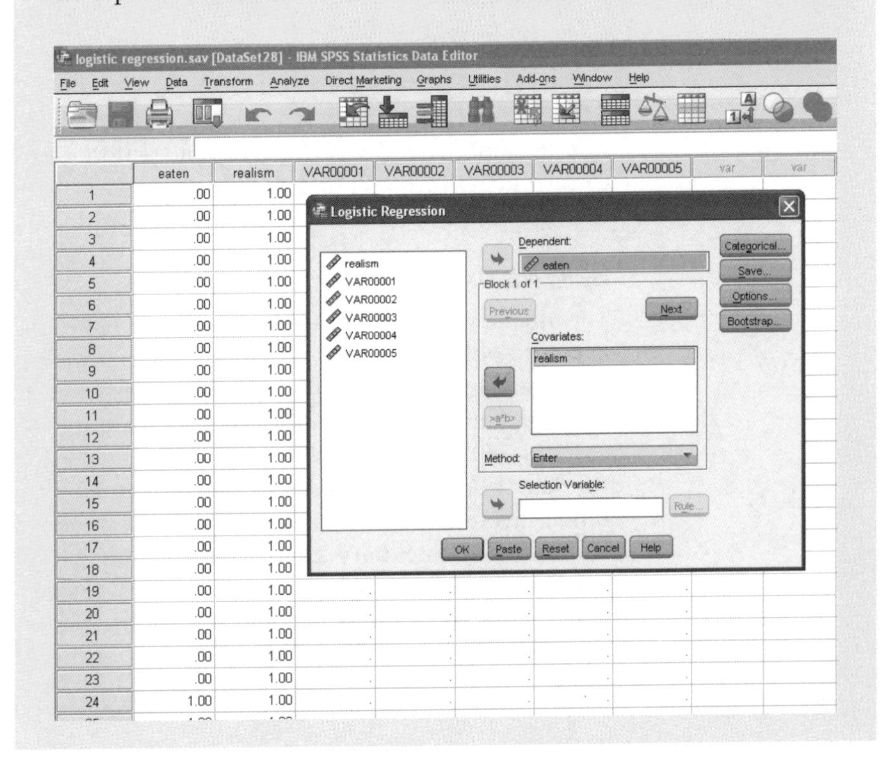

whether or not it was eaten (0 or 1). Since there are 150 models this involves entering 300 numbers. However, you can also do it much quicker in this case by weighting the data as in Example 7.1. Next, click on the **Analyse** menu, move onto the **Regression** bar and click on **Binary Logistic**. SPSS will produce the **Logistic Regression** dialogue box. Put the dependent variable (here **eaten**) into the **Dependent** box and the independent variable (here **realism**) into the **Covariate** box. The completed box, and the data screen are shown in previous page bottom.

Finally click on **OK**. SPSS will produce masses of output, of which the only useful bits, are the last two tables shown below.

Classification table[a]

Observed			Predicted: Eaten 0.00	Predicted: Eaten 1.00	Percentage Correct
Step 1	Eaten	0.00	64	23	73.6
		1.00	36	27	42.9
	Overall percentage				60.7

[a]The cut value is 0.500.

Variables in the equation

		B	S.E.	Wald	df	Sig.	Exp (B)
Step 1[a]	Realism	−0.417	0.207	4.045	1	0.044	.659
	Constant	0.501	0.438	1.306	1	0.253	1.651

[a]Variable(s) entered on step 1: realism.

The first table shows how well the logistic regression model predicts the numbers of eaten and uneaten models. The second table is like the table in a linear regression; it gives the size of the effect (B) of realism, which is just like the slope of a regression analysis and performs a test to see if B is significantly different from 0; in other words it tests whether the model has significantly improved the fit. Here $B = -0.417$ with a standard error of 0.207.

Using MINITAB

The usual way (and if you are using measurements the only way) of putting data into MINITAB is to enter each model separately in a large data sheet, with separate columns for the two characteristics: its realism (1, 2 or 3), and whether or not it was eaten (0 or 1). Since there are 150 models this involves entering 300 numbers. You could also do it by putting the frequencies into the data sheet, as in χ^2 for association but I don't recommend this as it gets rather involved. Next, click on the **Stat** menu, move onto the **Regression** bar and click on **Binary Logistic Regression**. MINITAB will produce the **Binary Logistic Regression** dialogue box. Put the dependent variable (here **eaten**) into the **Response** box and the independent variable (here **realism**) into the **Model** box. The completed box and the data screen are shown below.

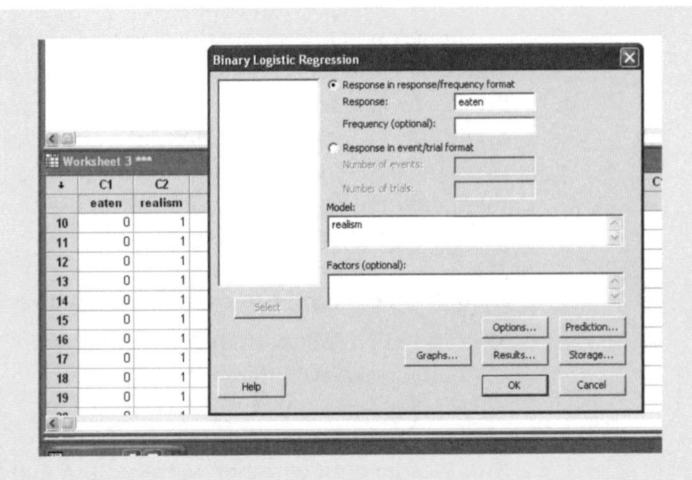

Finally click on **OK**. MINITAB will produce masses of output, of which the most useful bits are the following tables.

```
Logistic Regression Table

                                                       Odds     95% CI
Predictor        Coef   SE Coef      Z       P   Ratio  Lower  Upper
Constant     0.501132  0.438429   1.14   0.253
realism     -0.416578  0.207132  -2.01   0.044    0.66   0.44   0.99

Log-Likelihood = -99.976
Test that all slopes are zero: G = 4.135, DF = 1, P-Value = 0.042
Table of Observed and Expected Frequencies:
(See Hosmer-Lemeshow Test for the Pearson Chi-Square Statistic)
             Group
Value     1     2     3  Total
1
  Obs    17    19    27     63
  Exp  16.1  20.9  26.1
0
  Obs    33    31    23     87
  Exp  33.9  29.1  23.9
Total    50    50    50    150
```

The second table shows the observed and the expected numbers of eaten and uneaten models according to the model that the logistic regression has fitted. The first table is like the table in a linear regression; it gives the size of the effect (Coef) like the slope of a regression analysis and performs a test to see if Coef is significantly different from 0. In other words it tests whether the model has significantly improved the fit. Here Coef = −0.4166 with a standard error (SE Coef) of 0.207.

Step 3: Calculating the significance probability

SPSS and MINITAB both directly work out the probability (Sig. and P) that the slope would be different from 0.

Here Sig. = $P = 0.044$.

Step 4: Deciding whether to reject the null hypothesis

- If Sig. or $P \leq 0.05$ you should reject the null hypothesis. Therefore you can say that the factor has a significant effect on the frequencies of the two outcomes.
- If Sig. or $P > 0.05$ you have no evidence to reject the null hypothesis. Therefore you can say that the factor does not have a significant effect on the outcome.

Here Sig. = $P = 0.044 < 0.05$. Therefore we must reject the null hypothesis. We can say that the realism of the models significantly affects their probability of being eaten (the more realistic a model is the less likely it is to be eaten).

7.6.4 Logistic regression vs χ^2 tests

Since the data given in Example 7.4 is presented in a contingency table you might think that it could be analysed using a χ^2 test. This is certainly true for 2 × 2 contingency tables. In Example 7.4, however, the realism is ranked data (1 to 3), so the χ^2 test is inappropriate. In fact carrying out such a χ^2 test investigates whether the three types of model are different, but since it is not looking at the trend with increased realism, it comes out non-significant (P = 0.100). The great benefit of logistic regression is that it can investigate the effect of both ranked and ordinal data on binary outcomes. MINITAB will even allow you to calculate the effect of such factors on events with more than two possible outcomes.

7.6.5 Presenting the results of logistic regression

The results of a logistic regression are difficult to give graphically and so are better presented in text or by presenting the contingency table. For instance in this example you could give Table 7.2 and say

Table 7.2 The numbers of models eaten and left uneaten by the birds

	Unrealistic	Semi-camouflaged	Completely camouflaged	Total
Eaten	27	19	17	62
Uneaten	23	31	33	88
Total	50	50	50	150

Table 7.2 shows the numbers of the different types of models that were eaten or left uneaten. A logistic regression showed that better camouflaged models were eaten more rarely than less camouflaged ones ($\beta = -0.417$, $P = 0.044$).

7.7 Self-assessment problems

Problem 7.1

In an experiment to test the reactions of mice to a potential pheromone, they were run down a T-junction maze; the pheromone was released in one of the arms of the T. After the first 10 trials, 3 mice had turned towards the scent and 7 had turned away. After 100 trials, 34 had turned towards the scent and 66 had turned away. Is there any evidence of a reaction to the scent?

(a) After 10 trials
(b) After 100 trials

Problem 7.2

A cross was carried out between peas which were heterozygous in the two characters: height (tall H or short h) and pea colour (green G or yellow g). The following offspring were obtained.

	Number
Tall plants, green peas	87
Tall plants, yellow peas	34
Short plants, green peas	28
Short plants, yellow peas	11

For unlinked genes the expected ratios of each sort of plant are 9:3:3:1. Carry out a χ^2 test to determine whether there is any evidence of gene linkage between these characters.

Problem 7.3

A study of the incidence of a childhood illness in a small mining town showed that out of a population of 165 children, 9 had developed the disease. This compares with a rate of 3.5% in the country as a whole. Is there any evidence of a different rate in the town?

Problem 7.4

In a study of insect pollination, the numbers of insect visitors belonging to different taxonomic groups were investigated at flowers of different colours. The following results were obtained.

	Flower colour			Total
Insect visitors	**White**	**Yellow**	**Blue**	
Beetles	56	34	12	102
Flies	31	74	22	127
Bees and wasps	57	103	175	335
Total	144	211	209	564

(a) Carry out a χ^2 test to determine whether there is any association between the types of insects and the colour of the flowers they visit.
(b) Which cells have the three highest χ^2 values? What do these results tell you about the preferences of different insects?

Problem 7.5

A study was carried out to determine whether there is a link between the incidence of skin cancer and the possession of freckles. Of the 6045 people examined, 978 had freckles, of whom 33 had developed skin cancer. Of the remaining people without freckles, 95 had developed skin cancer. Is there any evidence that people with freckles have an increased risk of developing skin cancer?

Problem 7.6

A field study on the distribution of two species of newt found that of 745 ponds studied, 180 contained just smooth newts, 56 just palmate newts, 236 had both species present and the remainder had neither. Is there any association between the two species and, if so, what is it? χ^2

8 Designing experiments

8.1 Introduction

The role of experiments and surveys in biology is to help you answer questions about the natural world by testing hypotheses you have put forward. According to scientific papers (a highly stylised, and even misleading form of literature) this is amazingly simple. All you need do is to examine and take measurements on small numbers of organisms or cells. The results you obtain can then be subjected to statistical analysis to determine whether differences between groups or relationships between variables are real or could have occurred by chance. Nothing could appear simpler, but this is because scientific papers leave out the vast majority of work that scientists actually do, deciding on, designing and setting up their experiments.

Good scientists spend a lot of time making sure their experiments will work before they attempt them. Bad ones can waste a lot of time, effort and money carrying out badly designed experiments or surveys. Some experiments which they carry out could never work; in others the size of the sample is either too small to detect the sorts of effects which might be expected or much larger than necessary. Still other experiments are ruined by a confounding variable.

Before you carry out an experiment or survey, you should therefore ask yourself the following questions:

- Could it possibly work?
- How can I exclude confounding variables?
- How should I arrange things?
- How will I analyse the results?
- How large should my samples be?

Most of the work a scientist does is aimed at answering these questions and is carried out *before* the final experiment is performed. The key to success in experimentation is careful forward planning and preliminary work. The main techniques are preparation, replication, randomisation and blocking.

8.2 Preparation

To find out whether an experiment could work, you must first find out something about the system you are studying by reading the scientific literature, by carrying out the sort of rough calculations we examined in Chapter 2 or by making preliminary examinations yourself.

Example 8.1

Consider an experiment to test whether the lead from petrol fumes affects the growth of roadside plants. There would be no point in carrying out an experiment if we already knew from the literature that lead only reduces growth at concentrations above 250 ppm, while levels measured at our roadside were only 20 ppm.

8.3 Excluding confounding variables

In an experiment you are usually investigating the effect of one (or at most two) factors on organisms, organs or cells, while keeping all the other conditions the same. For instance if you want to investigate the effect of a drug treatment on rats, you should give it to one set of rats (the treated group) while treating a second group (the control group) exactly the same except that they are not given the drug. This should exclude potential **confounding variables** which might affect the two groups differently and mess up the experiment.

confounding variables Variables which invalidate an experiment if they are not taken into account.

Excluding confounding variables is not as easy as it might first seem, however. For instance if you are injecting the treated rats with the drug, you are not only giving them the drug, but also subjecting the rats to the ordeal of the injections. Therefore your control rats should also be injected, but with pure saline solution alone. This sort of intervention is known as a **procedural control**; you could have both normal and procedural controls for this sort of treatment.

Use of procedural controls is, of course, obligatory in human drug trials, in which control patients are treated exactly the same as ones who are given the test drug, except that they are given an inactive dose or **placebo**. Drug trials are also **double blind**; the researcher does not know which patients are given the drug, so that the possible confounding variable of their expectations are not included.

placebo A non-active control treatment used in drug trials.

8.4 Replication and pseudoreplication

replication The use of large numbers of measurements to allow one to estimate population parameters.

Everything we have done so far in this book has emphasised that you always need to repeat observations and to include **replication** in your experiments. However, this is not always as straightforward as it sounds. True **replicates** are individuals that have been subjected to exactly the same treatments as

replicates
The individual data points.

members of a different treatment, except for the one experimental treatment in which they differ. As we have seen, this helps exclude potential confounding variables.

Clearly this is straightforward for some factors. You can easily arrange an experiment so that the replicates are all kept at the same temperature, at the same light levels and over the same period of time. However, it is not possible to keep every factor the same; for instance, you cannot grow plants in exactly the same position as each other, or in exactly the same soil, and you cannot harvest and test them at exactly the same time. In such cases, where differences are inevitable, systematic errors between samples can be avoided by using **randomisation** techniques. You should randomise the position of plants and the order in which they are tested using random numbers from tables or computers, so that none of the groups is treated consistently differently from the others.

Failure to randomise can lead to **pseudoreplication**. You might, for instance, be interested in the effect of temperature on bacterial growth, but only have two growth rooms. You might then grow one set of plates in one growth room set at a low temperature and the other set in the other growth room at a higher temperature. The problem is that this introduces the confounding variable of the growth rooms; they might be different in other respects than just temperature. Therefore you would be testing whether there is a growth room effect, not a purely temperature effect. One way of getting round this is to periodically shift your plates between the rooms, swapping their temperature regimes. This method of controlling the confounding variable is not perfect, but it may be the best you can do with low resources.

Another example of pseudoreplication is if you take lots of measurements *on a single subject.* For instance if you are looking at the effect of fertiliser on the leaves of trees, you might grow two trees, one with high and one with low fertiliser, and take measurements on lots of leaves. You might think that each leaf was a replicate and use it to show that trees with high fertiliser levels grow better than ones with low levels. However, in reality you have only one replicate per treatment, the two trees! All you would be testing would be whether the leaves in tree 1 had leaves that were different from that of tree 2. To get over this problem, you should use several trees, and analyse the results on the leaves using **nested ANOVA** (Section 5.10).

Pseudoreplication is a particular problem in molecular biology. If you want to compare ribonucleic acid (RNA) from a tumour, for instance, with RNA from control tissue, you might take a single RNA extract from each and process them three times. These look at the effect of the technology on the variability and are known as **technical replicates**. Alternatively you could take three separate RNA replicates from each and process them just once. These look at the effect of biological variability and are known as **biological replicates**. Ideally, you need to have several biological replicates as well as several technical replicates in your experimental design, and to analyse the results with nested ANOVA.

8.5 Randomisation and blocking

blocking
A method of eliminating confounding variables by spreading replicates of the different samples evenly among different blocks.

Randomisation seems straightforward, but it can lead to a rather uneven spread of replicates. In a square field plot, for instance, you could get treaments bunched up into the corners or the centre (Figure 8.1a). A way of getting around this is to use what is known as a **Latin square** design. The plots are arranged so that each treatment appears once in each row, and once in each column (Figure 8.1b). Another problem would occur if you had a field plot that is very long and thin or split up between several fields; you might easily get different numbers of replicates of each sample at different ends or in different fields (Figure 8.2a). In this case you should split the plot into several blocks (Figure 8.2b) and randomise the same number of replicates from each sample within each block. The same should be done if testing or harvesting of your experiment takes several days. You should test an equal number of replicates from each sample each day. This use of **blocking** has the added advantage that, if you analyse your data using **two-way ANOVA**, you can also investigate separately the differences between the blocks.

(a)

A	B	C	B
A	C	A	B
D	C	B	C
D	A	D	D

(b)

A	B	C	D
B	A	D	C
C	D	B	A
D	C	A	B

Figure 8.1 The Latin square design helps avoid unwanted bunching of treatments. (a) A fully random design might result in all treatment A towards the top left. **(b)** In a Latin square only a single replicate of each treatment is arranged in each row and column.

(a)

Field 1

A	B	B	A
A	A	A	B

Field 2

B	B	A	A
B	B	A	B

(b)

Field 1

A	B	B	A
B	A	A	B

Field 2

A	A	B	A
B	B	A	B

Figure 8.2 Blocking can help to avoid confounding variables: an agricultural experiment with two treatments, each with eight replicates. (a) The treatments have been randomised, but this has produced an uneven distribution of treatments, both between the fields (more of treatment A is in field 1) and within them (in field 2 treatment B is concentrated at the left-hand side). **(b)** The treatments have been randomised within blocks (two applications of each treatment are made to each side of each field); this removes the possible confounding variable of position.

Example 8.2

An experiment was carried out to test the effect of a food supplement on the rate of growth of sheep. Individual sheep were kept in small pens and given identical amounts of feed, except that half the sheep were given the supplement and half were not. Because of space limitations, however, the sheep had to be split between three sheds, with 8 sheep in shed 1, 6 in shed 2 and 6 in shed 3. To ensure adequate blocking and take account of the possible effect of the sheds, the experimenters made sure that there were equal numbers of sheep that were given the supplement and sheep without the supplement in each shed. The sheep were kept for 6 months in these conditions before being weighed.

The following results were obtained:

Shed number	1	1	1	1	1	1	1	1	2	2
Treatment	–	–	–	–	+	+	+	+	–	–
Sheep mass (kg)	56	48	54	57	59	61	55	64	67	64

Shed number	2	2	2	2	3	3	3	3	3	3
Treatment	–	+	+	+	–	–	–	+	+	+
Sheep mass (kg)	59	65	68	65	45	53	50	57	53	56

Which statistical test should be used to analyse the results?

Solution

Since there are just two treatments and you are looking for *differences* between sheep that were given the supplement and those that were not, it looks at first glance like it might be best to analyse the results using a two-sample t test. However, because there is blocking between the sheds, you can use shed as a second factor giving six treatments in all. The test to use is therefore two-way ANOVA. Carrying out such a test in SPSS gives the following results:

Descriptive statistics

Dependent variable: shpweight

Shed	Supplement	Mean	Std. deviation	N
1.00	0.00	53.7500	4.03113	4
	1.00	59.7500	3.77492	4
	Total	56.7500	4.83292	8
2.00	0.00	63.3333	4.04145	3
	1.00	66.0000	1.73205	3
	Total	64.6667	3.14113	6
3.00	0.00	49.3333	4.04145	3
	1.00	55.3333	2.08167	3
	Total	52.3333	4.36654	6
Total	0.00	55.3000	6.86456	10
	1.00	60.3000	5.05635	10
	Total	57.8000	6.40395	20

Tests of between-subjects effects

Dependent variable: shpweight

Source	Type III sum of squares	df	Mean square	F	Sig.
Corrected model	607.700[a]	5	121.540	9.922	0.000
Intercept	65 867.045	1	65 867.045	5376.902	0.000
shed	471.033	2	235.517	19.226	0.000
supplement	117.333	1	117.333	9.578	0.008
shed * supplement	11.667	2	5.833	.476	0.631
Error	171.500	14	12.250		
Total	67 596.000	20			
Corrected total	779.200	19			

[a]R Squared = 0.780 (Adjusted R Squared = 0.701).

It can be seen that both shed and supplement have significant effects on sheep weight (Sig. = 0.000 and 0.008 respectively), and from the descriptive statistics two things can be seen:

1. Sheep in shed 2 were the heaviest and in shed 3 were the lightest.
2. Sheep given the supplement were heavier than control animals.

If a two-sample t test had been used it would have shown no significant effect of the supplement ($t = 1.855$, Sig. = 0.082) because the variability caused by the sheds swamps the small difference between sheep within each shed.

8.6 Choosing the statistical test

It is at this stage that you should use the decision chart (Figure 1.1) to work out which statistical test you are going to need to analyse your results. You should be clear about this. It will help ensure that you have chosen a sensible, straightforward experimental design. It will also help you decide how many replicates you will need in your experiment.

Here are some examples of deciding which tests to use following Figure 1.1.

Example 8.3

You are comparing the seed weights of 4 varieties of winter wheat, and you have weighed 50 seeds of each variety. What null hypothesis should you test and which test should you use?

Solution

If you are comparing seed weight (e.g. looking for differences) the null hypothesis is that there is no significant difference between the mean weight of each of the varieties. So which statistical test should you use?

- You have taken a *measurement* of weight for all 200 seeds, so you should choose the left-hand option.

- You are looking for *differences* between the varieties, so you should choose the left-hand option.
- You have *more than two* sets of measurements, so you should choose the right-hand option.
- You are investigating the effect of only *one factor*, variety, so you should choose the left-hand option.
- The measurements on the seeds are NOT in matched sets, so you should choose the right-hand option.
- You are examining a *continuous* measurement, weight, likely to be normally distributed and with a sample size of well over 20, so you should choose the boldface option.
- You should use the parametric test *one-way ANOVA*.

Example 8.4

You are investigating how the swimming speed of fish depends on their length, and you have measured both for 30 fish. What is your null hypothesis and which statistical test should you use to test it?

Solution

If you are investigating the relationship between swimming speed and length, the null hypothesis is that there is no significant association between them. So which statistical test should you use?

- You have taken *measurements* on speed and length, so you should choose the left-hand option.
- You are looking for a *relationship* between the two measurements, speed and length, so you should choose the right-hand option.
- One measurement, length, is clearly *unaffected by* the other, speed, so you should choose the right-hand option.
- Both length and speed are continuous measurements which are likely to be normally distributed; therefore you should carry out *regression analysis*.

Example 8.5

You are investigating the incidence of measles in children resident in hospitals and comparing it with the national average; you have surveyed 540 children. Which test should you use?

Solution

If you are comparing incidence of measles between hospitals and the national average the null hypothesis must be that there is no significant difference between them. So which statistical test should you use?

- You have counted the *frequencies* of children in two *categories* (with and without measles), so you should choose the right-hand option.
- You are comparing your sample with an *expected outcome* (the national average), so you should choose the left-hand option.
- You should calculate χ^2 for *differences*.

Example 8.6

You are investigating the relationship between the weight and social rank of domestic hens, and you have observed 34 birds. Which test should you use?

Solution

If you are investigating the relationship between weight and social rank, the null hypothesis is that there is no relationship between them. So which statistical test should you use?

- You have taken *measurements* on weight and assigned social rank to your birds, so you should choose the left-hand option.
- You are looking for a *relationship* between the two measurements, weight and social rank, so you should choose the right-hand option.
- The measurements are *not unaffected by* each other – weight can affect rank, and rank can also affect weight – so you should choose the left-hand option.
- Since social rank is by definition *rank* data you should choose the non-parametric option in normal type.
- You should carry out *rank correlation*.

8.7 Choosing the number of replicates: power calculations

Good experimental design is not enough if you don't have enough replicates in your experiments to detect effects that you are interested in. The key to working out *how many* replicates you need in your samples is knowledge of the system you are examining. You must have enough replicates to allow statistical analysis to tease apart the possible effects from random variability, but not many more or else you would be wasting your time. Therefore you need to know two things: the size of the effects you want to be able to detect; and the variability of your system. In general, the smaller the effect you want to detect and the greater the variability, the larger the sample sizes you will need. For parametric tests for differences it is relatively straightforward to work out how many replicates you need, using so-called **power calculations**. Nowadays, there are many programs (including MINITAB) that allow you to perform power calculations, but it is important to understand how they work, and how you can make rough power calculations yourself.

As we saw in Chapter 4, in t tests a difference becomes significant when t is around 2, in other words, when the mean is around two standard errors away from the expected value. In other words, when

$$D \approx 2\,\overline{SE}$$

where D is the difference and $\overline{SE}$ is the standard error. If the mean value of the population from which you took your sample was two standard errors away from the expected mean, you would therefore only be able to detect that there was a difference exactly half the time (Figure 8.3a) (because half the time the

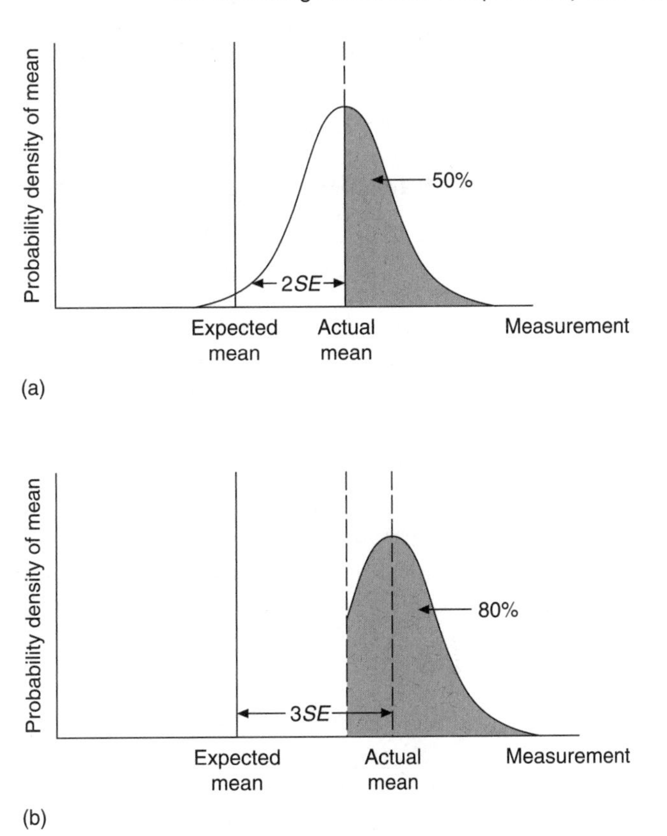

Figure 8.3 (a) An effect will be detected roughly 50% of the time if the expected value is two standard errors away from the actual population mean **(b)** To detect a significant difference between a sample and an expected value 80% of the time, the expected value should be around three standard errors away from the population mean.

sample mean would be higher than the population mean and half the time lower). To be able to detect that there was an 80% difference of the time (the accepted criterion for power calculations) the population mean would have to be approximately a further standard error away from the expected mean (Figure 8.3b). In other words

$$D \approx 3\,\overline{SE}$$

But standard error is the standard deviation divided by the square root of the number in the sample ($\overline{SE} = s/\sqrt{N}$). Therefore the smallest detectable difference, D is given by

$$D \approx 3s/\sqrt{N} \tag{8.1}$$

The number of replicates you need, N, to get an 80% chance of detecting the difference D can therefore be found by rearranging the equation, so that

$$N \approx 9(s/D)^2 \tag{8.2}$$

Before carrying out an experiment you should therefore strive to obtain figures for the size of the effect you want to detect and the likely standard deviation of samples. You could do this by looking up values from the literature or by carrying out a small pilot experiment. This will allow you to carry out power calculations, either by hand, or using MINITAB.

8.7.1 One-sample and paired *t* tests

Carrying out power calculations is best shown by looking at an example.

Example 8.7

A survey is to be carried out to determine whether workers on an oil platform have higher levels of stress hormone than do the general population. The mean and standard deviation for the general population are 2.15 (0.69) nM. Assuming the workers show the same variation as the general population, how many would be needed to test to detect

(a) A 20% (0.43 nM)
(b) A 10% (0.215 nM) difference in hormone concentration?
(c) The total number of workers available is 124. What difference would be detectable if they were all used?

Solution

Using a calculator

(a) Using equation 8.2, $n \approx 9 \times (0.69/0.43)^2 \approx 23$.
(b) $n \approx 9 \times (0.69/0.215)^2 \approx 93$.
(c) Using equation 8.1 $D \approx 3 \times 0.69/\sqrt{124} = 0.186$ nM.

Using MINITAB

Click on the **Stat** menu, move onto the **Power and Sample Size** bar and click on **1-Sample t.** MINITAB will produce the **Power and Sample Size for 1-Sample t** dialogue box. Put in the value for the **standard deviation** (here 0.69), give 0.8 as the **Power value,** and put in either the **Sample sizes:** or the **Differences:** MINITAB will calculate the other. For instance for (a) the completed dialogue box is shown below:

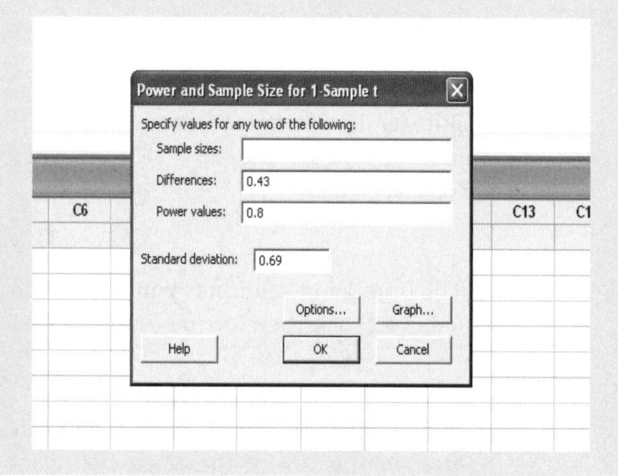

Finally click on **OK** to run the calculation. MINITAB will come up with the following output for (a).

Power and sample size

```
1-Sample t Test

Testing mean = null (versus not = null)

Calculating power for mean = null + difference

Alpha = 0.05  Assumed standard deviation = 0.69

              Sample   Target
Difference     Size    Power   Actual Power

      0.43       23      0.8       0.814903
                 ↑
```

For (a) MINITAB gives a value for the sample size of 23.
For (b) MINITAB gives a value for the sample size of 83.
For (c) MINITAB gives a value of 0.174967.
The rough calculations give a reasonable figure, but not exactly correct!

8.7.2 Two-sample *t* tests

The smallest detectable difference D in two-sample t tests also occurs when t is around 2. In this case, however, D is twice the standard error of the difference $\overline{SE}_d$, which itself is around 1.5 times the standard error of each sample. So in this case

$$D \approx 2\,\overline{SE}_d \approx 2(1.5\,\overline{SE}) \approx 3\,\overline{SE}$$

and using $\overline{SE} = s/\sqrt{N}$ as we did above, this gives the expression

$$D \approx 3s/\sqrt{N} \qquad (8.3)$$

To be able to detect that there was an 80% difference of the time (the accepted criterion for power calculations) the population mean would have to be approximately a further standard error away from the expected mean. In other words

$$D \approx 4\,\overline{SE}$$

But standard error is the standard deviation divided by the square root of the number in the sample ($\overline{SE} = s/\sqrt{N}$). Therefore the smallest detectable difference, D is given by

$$D \approx 4s/\sqrt{N} \qquad (8.4)$$

The number of replicates you need, N, to get an 80% chance of detecting the difference D can therefore be found by rearranging the equation, so that

$$N \approx 16(s/D)^2 \tag{8.5}$$

Therefore to detect a given smallest difference D with a two-sample t test, you need a larger value of N than with a one-sample t test.

Example 8.8

Tortoises from two different volcanic islands are to be compared. Assuming the standard deviation of each population's mass is 30 g, how many tortoises should be sampled from each island to detect (a) a 20 g difference and (b) a 10 g difference in mean mass?

Solution

Using a calculator

(a) $N \approx 16 \times (30/20)^2 \approx 36$.
(b) $N \approx 16 \times (30/10)^2 \approx 144$.

Using MINITAB

Do the same as for the one-sample t test but click on **2-Sample t** and fill in the boxes in the same way.

For (a) MINITAB gives a value for the sample sizes of 37.
For (b) MINITAB gives a value for the sample sizes of 143.

8.7.3 ANOVA

The results of power calculations for ANOVA are the same when one has only two samples. However, for more than two samples the sample sizes needed go up because one would be carrying out several post hoc tests and the significance probability for each has to be reduced.

Power calculations for ANOVA are best carried out in MINITAB by clicking on the **One-Way ANOVA** bar. The dialogue box is then filled in as for t tests except you also have to give the number of samples by filling in the **Number of levels:** box.

Example 8.9

The survey of tortoises in Example 8.8 is extended to end up with four volcanic islands. Repeat the power calculation s to work out sample sizes to detect (a) a 20 g difference and (b) a 10 g difference in mean mass.

Solution

Using MINITAB

Do the same as for the **2-Sample t** but put 4 into the **Number of levels** box and fill in the boxes in the same way.

For (a) MINITAB gives a value for the sample sizes of 51.
For (b) MINITAB gives a value for the sample sizes of 198.

8.7.4 Chi-squared tests for differences

If you have an expected outcome for the frequency of a particular category, it is once again straightforward to work out the sample size you require. As we saw in Chapter 7, the standard error s for the proportion with which a particular category will occur is given by the expression

$$s = (1 - p)/(N - 1) \tag{8.6}$$

Where p is the expected proportion. The smallest proportional difference d that will be statistically significant is around twice the standard error, so

$$d \approx 2\sqrt{p(1 - p)/(N - 1)} \tag{8.7}$$

To get a significant result 80% of the time one would have to have a difference of a further standard error, so that

$$d \approx 3\sqrt{p(1 - p)/(N - 1)} \tag{8.8}$$

Rearranging 8.8 gives an expression for the sample size N that would be required to detect a proportional difference, d:

$$N \approx [9p(1 - p)/d^2] + 1 \tag{8.9}$$

Example 8.10

We want to know how many people in a town we need to test to detect whether there is a 1% difference in the incidence of hair lice from the national figure of 4%.

Solution

Inserting values for p of 0.04 and for d of 0.01 into the equation, we get $N \approx [9 \times 0.04 \times 0.96/0.01^2] + 1 \approx 3456$.

8.7.5 Tests for relationships

It is also straightforward to determine before you have done the experiment the sample size you need to test for **relationships**. Once again, the larger the sample you take the smaller the association you will be able to detect. For instance if you are carrying out correlation analysis, the minimum correlation coefficient that you will be able to detect is simply the critical value for $N - 2$ degrees of freedom, where N is the sample size.

Example 8.11

In an investigation of the relationship between the size of people's hands and the length of their feet, 40 students were measured. What is the minimum correlation coefficient that could be detected? What sample size would be needed to detect a correlation coefficent of 0.2?

Solution

With 40 students, looking up the critical value of r for $40 - 2 = 38$ degrees of freedom reveals a critical value of r around 0.312. To detect a correlation coefficient of 0.2, the number of degrees of freedom needed is around 95, meaning a sample size of around 97 (say 100 to be on the safe side) would be needed.

You can also use this technique to determine the sample size needed for regression analysis.

8.8 Dealing with your results

During your experiment you should look at your results as soon as possible, preferably while you are collecting them, and try to think what they are telling you about the natural world. Calculate the mean and standard deviations of measurements, plot the results on a graph or look at the frequencies in different categories. Once you can see what seems to be happening, you should write down your ideas in your laboratory book, think about them and then tell your supervisor or a colleague. Do not put your results into a spreadsheet and forget about them until the write-up.

Only after you have worked out what you think is happening should you carry out your statistical analysis to see if the trends you identified are significant. Usually, if a trend is not obvious to the naked eye it is unlikely to be significant. So always use statistics as a tool; do not allow it to become the master!

8.9 Self-assessment problems

Problem 8.1

A clinician wants to find out if there is any link between energy intake (in calories) and heart rate in old people. She collects data on both of them from 150 volunteers.

What is her null hypothesis? And which statistical test should she choose to determine whether energy intake and heart rate are linked?

Problem 8.2

An ecologist collects data about the numbers of individuals that belong to five species of crow feeding in three different habitats: farmland, woodland and mountain. He wants to determine whether different crows are distributed non-randomly in different habitats.

What is his null hypothesis and how will he analyse his data to determine whether different crows are distributed non-randomly in different habitats?

Problem 8.3

A doctor wants to find out if there is any difference in insulin levels between three races of people: Afro-Caribbean, Asian and Caucasian. He collects data on insulin levels from 30 people of each race.

What is the null hypothesis and which statistical test should he use to answer his question?

Problem 8.4

A genetics student wants to find out whether two genes are linked: one for shell colour (brown dominant, yellow recessive) and one for having a banded or plain shell (banded dominant, plain recessive). To do this, she crosses two pure-bred lines of brown-banded snails with yellow plain snails. The result is an F1 generation, all of which are brown and banded. These are crossed to produce an F2 generation.

What is the null hypothesis and which statistical test should she perform to test whether there is in fact any linkage?

Problem 8.5

An ecologist wants to find out whether the levels of pesticide residue found in kestrels differ at different times of the year. She measures pesticide levels in ten birds, repeating the measurements on each bird every 2 months.

What is the null hypothesis and which statistical test should she perform to test whether pesticide levels are different at different times of the year?

Problem 8.6

A new medication to lower blood pressure is being tested in field trials. Forty patients were tested before and after taking the drug. Which test should the clinicians use to best determine whether it is having an effect?

Problem 8.7

It has been suggested that pot plants can help the survival of patients in intensive care by providing the room with increased levels of oxygen from photosynthesis. Carry out a rough calculation to work out if this idea is worth testing. (*Hint*: estimate how fast the plant is growing and hence laying down carbohydrates and exporting oxygen.)

Problem 8.8

In an investigation into starch metabolism in mutant potatoes, the effect of deleting a gene is investigated. It is expected that this will reduce the level to which starch builds up. Large numbers of previous experiments have shown that the mean level of starch in ordinary potatoes is 21 M with standard deviation 7.9 M. Assuming the standard deviation in mutants is similar to that in ordinary potatoes, how many replicates would have to be examined to detect a significant difference in mutants whose mean starch level is 16 M, a 5 M fall?

Problem 8.9

An experiment is being designed to test the effect of shaking maize plants on their extension growth. There will be two groups: shaken plants and unshaken controls. It is known that maize usually has a mean height of 1.78 m with standard deviation of 0.36 m. How many replicates of experimental and control plants must be grown to detect a height difference of 0.25 m?

Problem 8.10

The national rate of breast cancer is given as 3.5% of women over 45 years old. It has been suggested that silicone implants may increase the rate of such cancers. How many women with implants would need to be tested to detect a doubling of the risk?

Problem 8.11

Design an experiment to test the relative effects of applying four different amounts of nitrogen fertiliser, 0, 3.5, 7 and 14 g of nitrogen per square metre per year, in 25 fortnightly applications (researchers get a fortnight off at Christmas) on to chalk grassland at Wardlaw, Derbyshire. The field site is split into 16 plots, each of dimensions 1 m $\times$ 1 m in an 8 m $\times$ 2 m grid (see grid). You are supplied with a quantity of 20×10^{-3} M ammonium nitrate solution fertiliser. Mark the different plots and describe exactly what you would apply to each plot.

9 More complex statistical analysis

9.1 Introduction to complex statistics

So far we have investigated how to analyse fairly simple experiments and surveys: ones in which you are looking at how one (or at most two) factors affect another, or at the relationships between two sets of measurements or factors. I would strongly advise most working biologists to limit themselves as far as they can to such simple experiments and to the statistical tests in the decision chart. Quite often teachers of statistics find that inexperienced scientists (or just ignorant ones!) who so often come to us at the end of their projects to 'help with their stats' have designed their experiments so poorly that they are not properly controlled, or there is a confounding variable. This means they have to be analysed using more complicated statistics that they don't really understand. The moral is always to think carefully about how your experiments will be analysed *before* carrying them out.

However, there are legitimate cases in which biologists may want or need to do rather more complex things and use more complicated statistical analyses.

- To save time you may want to design experiments that investigate several factors at once.
- You may not be able to control all the conditions in your experiments, so you may need to be able to carry out statistical analysis that makes allowances for this and mixes the two types of analysis: comparing groups and looking at relationships.
- You may want to investigate the relationships between several sets of measurements you have taken in a survey.
- You may have taken large numbers of measurements on a single group of organisms or cells and be interested in exploring the data. In particular, you might want to see whether the measurements allow the groups to be split up into several subgroups, and which ones are most similar.

This chapter introduces you to the main types of complex statistical analysis (none of which I have myself ever used!).

9.2 Experiments investigating several factors

In Chapter 5 we saw how you can use two-way ANOVA and nested ANOVA to analyse complex experiments in which two factors can be investigated simultaneously. It is also possible to examine the effect of three or more factors using three or more way ANOVA. Such tests can readily be performed in both SPSS and MINITAB using the same techniques as we used for two-way ANOVA. Of course you need to be careful to distinguish between the main factors and nested factors, but the principles are the same, and the data is analysed within the **General Linear Model** dialogue boxes.

There are a few problems you have to remember if you want to perform such complex analyses. First, if you want to perform a complete factorial experiment, and you have three or more factors the number of treatments you will have to set up starts to rise alarmingly. For instance, if you have three factors, each of which is found in two states you will have $2 \times 2 \times 2 = 8$ treatments. If each treatment is found in three states there will be 27! Second, the analysis will not only give you results about the main effects of the three factors but also the interactions between them. For a three-way ANOVA there will be three interaction terms between each pair of factors, and one extra one for the interaction of all three factors. These interaction terms become increasingly numerous and difficult to understand. Finally, all these analyses assume that the data is normally distributed. At present there are no straightforward non-parametric tests that can simultaneously test for several factors.

All of the tests for differences we have examined so far investigate the effect of factors on just a single measurement or variable (such as yield or size). They are what is known as **univariate** tests, so all the ANOVA tests we have examined are examples of **univariate ANOVA**. Usually biologists carry out this sort of analysis even when they want to examine the effect of the treatments on several different measurements. They simply carry out two or more ANOVAs. However, a theoretically better way of carrying out such analyses is by performing a **multivariate analysis of variance (MANOVA)** which investigates to see whether there are differences in a combination of measurements. This has the advantage that it might help find significant differences between groups when the effects on a number of individual measurements are all close to being significant. It also allows you to investigate how the various outcome measurements are related. This is a complex procedure, though, and it is hard to interpret what the results of a MANOVA actually mean in physical terms. For this reason, this technique is more often used by psychologists than biologists, so those interested are recommended to look up the theory and practice in Field (2009).

9.3 Experiments in which you cannot control all the variables

Sometimes you may not be able to control all the variables within your experiments. For instance you might have taken measurements of bone density in two groups of women, one of which was put under an exercise regime, the other of which was not. However, in both groups you might have women of widely varying age, which also potentially affects women's bone density. Consequently, age might act as a confounding variable, preventing you from finding a significant result if you simply analysed the experiment using a two-sample t test or one-way ANOVA.

Fortunately, you can make use of the fact that analysis of variance and regression are actually both manifestations of a single analytical technique. They both operate by apportioning and comparing the variability due to a possible effect (of one or more factors in ANOVA and one or more variables in regression) with the residual variability. The two techniques can therefore be combined (as they are in SPSS and MINITAB) within the **general linear model (GLM)**.

general linear model (GLM)
A series of tests which combine ANOVA and regression analysis to allow powerful analysis of complex data sets.

Analysis of covariance (ANCOVA) is the most commonly used example of a technique within GLM; it allows you to combine the effect of a single factor with a single variable. In the example of women's bone density, for instance, analysing the data using ANCOVA allows you to do two things.

ANCOVA
Abbreviation for analysis of covariance: a series of tests that determines whether there are significant differences between groups, while accounting for variability in other measurements.

1. To determine whether calcium affects bone density while allowing for the effects of age.
2. To determine whether age affects bone density, while allowing for the effects of the calcium treatment.

9.3.1 Carrying out the test

The test is best introduced through the example of women's bone density as shown below.

Example 9.1

In an investigation of the effects of an exercise treatment on bone density in women, ten women were put onto an exercise programme, while ten were used as controls. After a year their bone density and age were both recorded and the following results obtained.

Controls	**Age**	56	50	69	52	72	58	42	80	62	68
	Density	54	62	45	64	48	56	63	42	56	51
Exercise treated	**Age**	50	59	75	65	55	60	57	54	67	61
	Density	67	56	47	65	76	59	63	64	49	65

Did the exercise alter their bone density?

Solution

Step 1: Formulating the null hypothesis

The null hypothesis is that exercise had no effect on bone density. ANCOVA also tests the other null hypothesis that age had no effect on bone density.

Step 2: Calculating the test statistic

Using SPSS

First put the data into three columns, say **age, treatment** and **density**. Next, click on the **General Linear Model** menu, move onto the bar and click on **Univariate**. SPSS will produce the **Univariate** dialogue box. Put the dependent variable (here **density**) into the **Dependent Variable** box, **treatment** into the **Fixed Factor(s)** box and **age** into the **Covariate(s)** box. Finally, click on **Options** and tick **Descriptive statistics.** The completed boxes and the data screen are shown below.

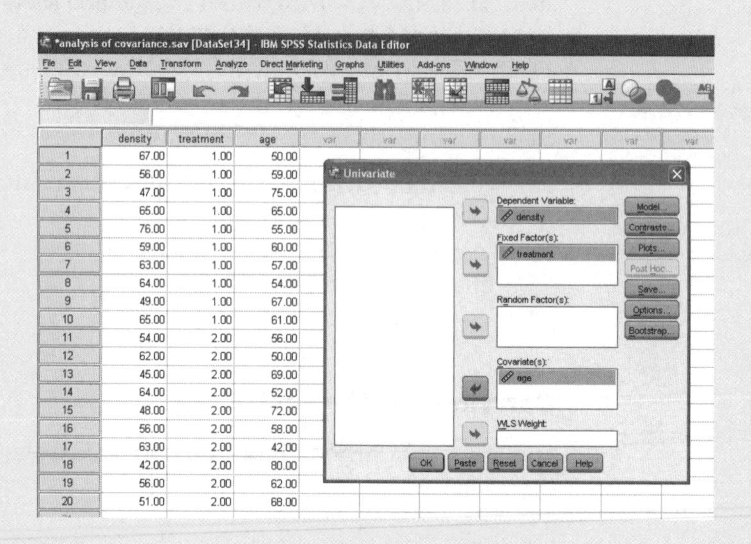

Finally click on **OK**. SPSS will produce an output of which the most useful part is shown in the following table:

Tests of between-subjects Effects

Dependent variable:density

Source	Type III sum of squares	df	Mean Square	F	Sig.
Corrected model	1 060.010[a]	2	530.005	23.538	0.000
Intercept	4430.254	1	4 430.254	196.751	0.000
age	815.010	1	815.010	36.195	0.000 ←
treatment	216.246	1	216.246	9.604	0.007 ←
Error	382.790	17	22.517		
Total	67 798.000	20			
Corrected total	1 442.800	19			

[a] R squared = 0.735 (adjusted R squared = 0.703).

This table gives F ratios for the effect of the treatment (9.604) and age (36.195).

Using MINITAB

First put the data into three columns say **age, treatment** and **density**. Next, click on the **Stat** menu, move onto the **ANOVA** bar and click on **General Linear Model**. MINITAB will produce the **General Linear Model** dialogue box. Put the dependent variable (here **Density**) into the **Responses** box and **Treatment** into the **Model** box. Then click onto the **Covariates** button and put **Age** into the **Covariates** box. The completed boxes and the data screen are shown below.

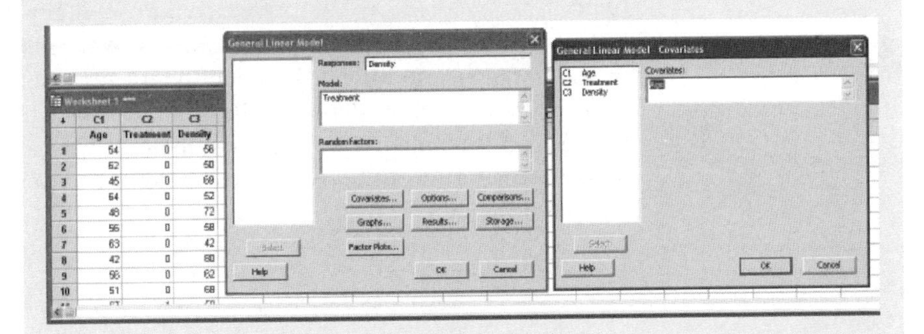

Finally click on **OK** in both boxes. MINITAB will produce masses of output, of which the most useful bits are the following tables.

Analysis of Variance for Density, using Adjusted SS for Tests

Source	DF	Seq SS	Adj SS	Adj MS	F	P
Age	1	973.59	1131.54	1131.54	36.20	0.000
Treatment	1	159.75	159.75	159.75	5.11	0.037
Error	17	531.46	531.46	31.26		
Total	19	1664.80				

S = 5.59126   R-Sq = 68.08%   R-Sq(adj) = 64.32%

Term	Coef	SE Coef	T	P
Constant	116.584	9.389	12.42	0.000
Age	-0.9719	0.1616	-6.02	0.000

This table gives F ratios for the effect t of the treatment (5.11) and age (36.20) and presents the result of a t test to compare the slope of the regression line of density vs age with 0. Here $T = -6.02$.

Step 3: Calculating the significance probability

SPSS and MINITAB have both directly worked out the probability that the means of the density are the same, accounting for the effect of treatment.

Here Sig. = 0.007 and $P = 0.037$.

Step 4: Deciding whether to reject the null hypothesis

- If Sig. or $P \leq 0.05$ you should reject the null hypothesis. Therefore you can say that the treatment has a significant effect.
- If Sig. or $P > 0.05$ you have no evidence to reject the null hypothesis. Therefore you can say that the treatment has no significant effect.

Here Sig. = 0.007 and $P = 0.037 < 0.05$. Therefore we must reject the null hypothesis. We can say that the exercise treatment has a significant effect on bone density: the treatment increases it. There are two things to note. First SPSS and MINITAB give different results! Second, the analysis also shows that bone density falls significantly with increasing age. If we had not included it in the analysis and just carried out a one-way ANOVA, the result of the treatment would not have come out significant (Sig. = 0.071).

9.4 Investigating the relationships between several variables

In Chapters 6 and 7 we saw how you can use correlation, regression, the x^2 test and logistic regression to investigate the relationships between two sets of variables. However, biologists often want to investigate the relationships between several sets of measurements. Once again there are several different ways of doing this, and the method you choose will depend on the type of data and the ways in which the different variables could influence each other.

correlation
A statistical test which determines whether there is linear association between two sets of measurements.

Correlation is the simplest method of investigating the relationships between more than two variables, where each variable could conceivably affect the others and vice-versa. All you need to do is to make all the possible comparisons you could do, using **Pearson correlation** for normally distributed data and **rank correlation** for non-normally distributed data or ranks. The result will be a large table or correlation coefficients that you can look through for significant correlations. There are two main problems with this simple approach. The first is that the more correlations you do the more likely you are to get significant correlations just by chance. You can get over this problem by decreasing the probability at which you consider significance to be reached. I recommend using the Dunn–Sidak method in which you reduce the probability according to the equation $P = 1 - (0.95)^{1/c}$ where k is the number of comparisons.

The second problem is not so easy to get over. Many of the relationships you uncover may not represent real causal relationships. Two variables may be correlated, not because there is a causal relationship between them but because one is co-correlated with a third variable that does have such a causal relationship. For instance many measurements on people may be correlated because they all change with age. However, this does not necessarily mean they affect each other. To get over this problem, the technique to use is **partial correlation**. This produces a measure of the correlation between two variables, when all the other variables are held constant. The technique is easy to carry out in SPSS (but not MINITAB!). Let's investigate the data about heart rate and blood pressure we examined in Example 6.1. The two variables were found to be closely correlated with a correlation coefficient of 0.860. What about adding a third measure, artery stiffness? To investigate how the three variables are related you can carry out several partial correlations.

For instance to see if the relationship between heart rate and pressure would remain significant, even if artery stiffness was taken into account, go into the **Correlate** menu, and click on **Partial Correlation**. SPSS will produce the **Partial Correlation** dialogue box. Put the two variables whose relationship you want to investigate (here **heartrate** and **pressure**) into the **Variables** box, and the variable for which you are controlling (here **artery**) into the **Controlling for** box. The completed boxes and the data screen are shown below.

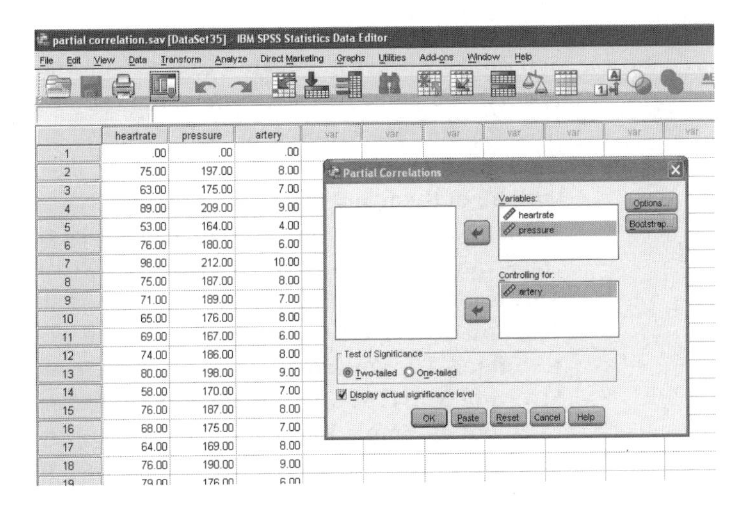

Finally click on **OK**. SPSS will produce the following table:

Correlations

Control variables			Heart rate	Pressure
Artery	Heartrate	Correlation	1.000	0.822 ←
		Significance (2-tailed)	.	0.000 ←
		df	0	27
	Pressure	Correlation	0.822	1.000
		Significance (2-tailed)	0.000	.
		df	27	0

This shows that even taking into account artery stiffness, heart rate and blood pressure are still strongly positively correlated.

Multiple regression allows you to determine how a single 'output' measurement is affected by several independent variables or cofactors. For instance you might be interested in seeing how the metabolic rate of patients is related to the levels of several hormones. If each hormone is measured in each patient, along with the metabolic rate, multiple regression allows you to work out if any of the hormones have a significant effect on metabolic rate. If so it will enable you to determine the line of best fit between hormone levels and metabolic rate, and tell you how much of the variability is accounted for by each hormone.

The main problem with using multiple regression is that there are several different ways in which it can be performed, and each can give different results! You can involve all the variables. Another way is to build them up *forwards*, starting with the variable that explains the most variability and continue until the last significant variable is added. Alternatively you can start with all of them and move *backwards* removing the ones with the least effect. A final way, *stepwise* adds new variables, like the forwards strategy but at any stage can add or take away any of the existing variables in the regression. This is a more flexible strategy, but none is absolutely guaranteed to get the best possible regression, so you are advised to proceed with caution with this technique. The advantage is that all of these techniques are supported in both SPSS and MINITAB (within the **Regression** menus) and they are as easy to carry out as simple linear regression. You just put in all the possible factors into the **Independent** or **Predictor** boxes and choose the technique to use from a menu.

logistic regression
A statistical test which analyses how a binary outcome is affected by other numerical characteristics

Logistic regression allows you to investigate how several variables or character states affect a binary output. Thus you might look at how several factors influence whether an animal is eaten or not, or which factors influence whether people smoke. The factors can be a mixture of character states, ranks and measurements.

9.5 Exploring data to investigate groupings

Finally, there are several techniques that can be used to investigate the relationships between several variables taken on a single group of subjects. Some techniques can help to split up the individuals as much as possible on the basis of differences in a combination of the variables. Other techniques can be used if you have already defined your groups. They can see which variables best *separate* the groups. Finally, yet others find out how *closely* the different subjects are related and work a dendrogram or family tree showing those relationships. These techniques are useful, but they should be mainly used as the basis for further research and to develop hypotheses which should then be tested if possible by more rigorous experiments or surveys.

9.5.1 Separating individuals

The two techniques that are most commonly used to do this are **principal component analysis (PCA)** and **factor analysis**. They do this by finding the combination of all the factors that separate the individuals the most (much as a line through a two-dimensional scatterplot is a combination of just two factors) and the other combinations that separate the individuals yet more. The points can then be plotted on a graph of the first factor against the second (or even a third if you are prepared to draw three-dimensional graphs! Inspecting the weighting of the factors shows which variables contribute most to the differences between the individuals.

9.5.2 Separating groups

The two techniques most commonly used to separate individuals which have already been split up on the basis of taxonomy or ecology are **canonical variate analysis (CVA)** and **discrimination function analysis**. The analysis is similar to PCA except that the variables are weighted so as to maximise the differences between the groups. Discriminate function analysis has the added feature that it allows you to determine how confidently new individuals can be assigned to the different groups.

9.5.3 Classifying and investigating similarities

The best way to determine similarities between individuals and to classify them into family trees or dendrograms is to use **cluster analysis**. These techniques are particularly useful for analysis of DNA sequences, but are also heavily used to investigate the phylogeny of organisms and the relationship between vegetation types.

9.5.4 A warning

Data exploration techniques can seem very attractive, as they allow the scientist to chuck huge amounts of 'data' into an analysis and get lots of information out. This can allow you to 'see the wood for the trees'. However, the problem is that the results they produce are hard to understand and it can be hard to see how reliable they are. Consequently, they should rarely be used as an end in themselves but should be used mainly to generate hypotheses, which can be subsequently tested by experiment. In science you can't find things out about your subjects before you have a reasonably good understanding about them and just collecting data is no substitute for thinking. For this reason, this book doesn't go deeply into this subject and the techniques are relevant mostly to particular areas of biology: morphometrics, systematics, community ecology and bioinformatics. Students of those subjects will no doubt find useful texts pointing them to the relevant techniques for their subject.

10 Dealing with measurements and units

10.1 Introduction

It is surprising, considering that most students of biology have studied mathematics for many years, how often they make errors in the ways they deal with and present numerical information. In fact there are many ways of getting things wrong. Primary data can be measured wrongly, or given too high or low a degree of precision. The data can be taken and presented in non-SI units, or mistakes can be made while attempting to convert to **SI** units. Calculations based on primary data can be carried out incorrectly. Finally, the answers can be given to the wrong degree of precision, in the wrong units, or with no units at all!

SI
Systéme International: the common standard of units used in modern science based on the metre, second and kilogram.

Many of the errors are made not only through ignorance but also because of haste, lack of care or even panic. This chapter shows how you can avoid such mistakes by carrying out the following logical sequence of steps carefully and in the right order: measuring, converting data into SI units, combining data together and expressing the answer in SI units to the correct degree of precision. It also gives some useful tables which can be consulted at any time for reference.

10.2 Measuring

Measurements should always be taken to the highest possible degree of precision. This is straightforward with modern digital devices, but it is more difficult in the more old-fashioned devices, which have a graduated analogue scale. The highest degree of precision of analogue instruments is usually to the smallest graduation of the scale. Using a 30-cm ruler, lengths can only be measured to the nearest millimetre. However, if the graduations are far enough apart, as they are on some thermometers, it is usually possible to judge the measurements to the next decimal place. This is made even easier by devices, like calipers or microscope stages, which have a vernier scale.

10.3 Converting to SI units

10.3.1 SI units

Before carrying out any further manipulation of data or expressing it, the data should be converted into the correct SI units. The *Système International d'Unités* (SI) is the accepted scientific convention for measuring physical quantities, under which the most basic units of length, mass and time are kilograms, metres and seconds respectively. The complete list of the basic SI units is given in Table 10.1.

All other units are derived from these basic units. For instance, volume should be expressed in cubic metres or m^3. Similarly density, mass per unit volume, should be expressed in kilograms per cubic metre or kg m^{-3}. Some important derived units have their own names; the unit of force (kg m s^{-2}) is called a newton (N), and the unit of pressure (N m^{-2}) is called a pascal (Pa). A list of important derived SI units is given in Table 10.2.

10.3.2 Dealing with large and small numbers

The problem with using a standard system, like the SI system, is that the units may not always be convenient. The mass of organisms ranges from 0.000 000 000 1 kg for algae to 100 000 kg for whales. For convenience, therefore, two different systems can be used to present large and small measurements. Both these systems also have the added advantage that large numbers can both be written without using a large number of zeros, which would imply an unrealistic degree of precision. It would be difficult to weigh a whale to the nearest kilogram (and pointless, since the weight will fluctuate wildly at this degree of precision), which is what the weight of 100 000 kg implies.

Use of prefixes

Using prefixes, each of which stands for a multiplication factor of 1000 (Table 10.3), is the simplest way to present large or small measurements. Any quantity can

Table 10.1 The base and supplementary SI units

Measured quantity	SI unit	Symbol
Base		
Length	metre	m
Mass	kilogram	kg
Time	second	s
Amount of substance	mole	mol
Temperature	kelvin	K
Electric current	ampere	A
Luminous intensity	candela	cd
Supplementary		
Plane angle	radian	rad
Solid angle	steradian	sr

Table 10.2 Important derived SI units

Measured quantity	Name of unit	Symbol	Definitions
Mechanics			
Force	newton	N	kg m s^{-2}
Energy	joule	J	N m
Power	watt	W	J s^{-1}
Pressure	pascal	Pa	N m^{-2}
Electricity			
Charge	coulomb	C	A s
Potential difference	volt	V	J C^{-1}
Resistance	ohm	Ω	V A^{-1}
Conductance	siemens	S	Ω^{-1}
Capacitance	farad	F	C V^{-1}
Light			
Luminous flux	lumen	lm	cd sr^{-1}
Illumination	lux	lx	lm m^{-2}
Others			
Frequency	hertz	Hz	s^{-1}
Radioactivity	becquerel	Bq	s^{-1}
Enzyme activity	katal	kat	mol substrate s^{-1}

Table 10.3 Prefixes used in SI

Small numbers						
Multiple	10^{-3}	10^{-6}	10^{-9}	10^{-12}	10^{-15}	10^{-18}
Prefix	milli	micro	nano	pico	femto	atto
Symbol	m	μ	n	p	f	a
Large numbers						
Multiple	10^{3}	10^{6}	10^{9}	10^{12}	10^{15}	10^{18}
Prefix	kilo	mega	giga	tera	peta	exa
Symbol	k	M	G	T	P	E

be simply presented as a number between 0.1 and 1000 multiplied by a suitable **prefix**. For instance, 123 000 J is better presented as 123 kJ or 0.123 MJ. Similarly, 0.000 012 m is better presented as 12 μm (not 0.012 mm).

prefix
A multiple or divisor of 1000 which allows large or small numbers to be readily expressed.

Use of scientific notation

The problem with using prefixes is that they are rather tricky to combine mathematically when carrying out calculations. For this reason, when performing a calculation it is usually better to express your data using **scientific notation**. As we shall see, this makes calculations much easier.

scientific notation
A method of representing large or small numbers, giving them as a number between 1 and 10 multiplied by a power of 10.

Any quantity can be expressed as a number between 1 and 10 multiplied by a power of 10 (also called an exponent). For instance, 123 is equal to 1.23 multiplied by 10 squared or 10^2. Here the exponent is 2, so it can be written as 1.23×10^2. Similarly, 0.001 23 is equal to 1.23 multiplied by the inverse of 10 cubed, or 10^{-3}. Therefore it is best written as 1.23×10^{-3}. And 1.23 itself is equal to 1.23 multiplied by 10 to the power 0, so it does not need an exponent.

A simple way of determining the value of the exponent is to count the number of digits from the decimal point to the right of the first significant figure. For instance, in 18 000 there are four figures to the right of the 1, which is the first significant figure, so $18\,000 = 1.8 \times 10^4$. Similarly, in 0.000 000 18 there are seven figures between the point and the right of the 1, so $0.000\,000\,18 = 1.8 \times 10^{-7}$.

Prefixes can readily be converted to exponents, since each prefix differs by a factor of 1000 or 10^3 (Table 10.3). The pressure 4.6 MPa equals 4.6×10^6 Pa, and 46 MPa equals $4.6 \times 10^1 \times 10^6 = 4.63 \times 10^7$ Pa.

10.3.3 Converting from non-SI units

Very often textbooks and papers, especially old ones, present quantities in non-SI units, and old apparatus may also be calibrated in non-SI units. Before carrying out calculations, you will need to convert them to SI units. Fortunately, this is very straightforward.

Non-SI metric units

The most common non-SI units are those which are metric but based on obsolete systems. The most useful biological examples are given in Table 10.4, along with a conversion factor. These units are very easy to convert into the SI system. Simply multiply your measurement by the conversion factor.

Example 10.1

Give the following in SI units:

(a) 24 ha
(b) 25 cm

Solution

(a) 24 ha equals $24 \times 10^4\ m^2 = 2.4 \times 10^5\ m^2$
(b) 25 cm equals $25 \times 10^{-2}\ m = 2.5 \times 10^{-1}\ m$

Litres and concentrations

The most important example of a unit which is still widely used even though it does not fit into the SI system is the litre (1 dm^3 or $10^{-3}\ m^3$), which is used in the derivation of the concentration of solutions. For instance, if 1 l contains 2 moles of a substance then its concentration is given as 2 *M* or molar.

The mole is now a bona fide SI unit, but it too was derived before the SI system was developed, since it was originally the amount of a substance which contains the same number of particles as 1 g (rather than the SI kilogram) of

Table 10.4 Conversion factors from obsolete units to SI[a]

Quantity	Old unit/symbol	SI unit/symbol	Conversion factor
Length	angstrom/Å	metre/m	1×10^{-10}
	yard	metre/m	0.9144
	foot	metre/m	0.3048
	inch	metre/m	2.54×10^{-2}
Area	hectare/ha	square metre/m^2	1×10^4
	acre	square metre/m^2	4.047×10^3
	square foot/ft^2	square metre/m^2	9.290×10^{-2}
	square inch/in^2	square metre/m^2	6.452×10^{-4}
Volume	litre/l	cubic metre/m^3	1×10^{-3}
	cubic foot/ft^3	cubic metre/m^3	2.832×10^{-2}
	cubic inch/in^3	cubic metre/m^3	1.639×10^{-5}
	UK pint/pt	cubic metre/m^3	5.683×10^{-4}
	US pint/liq pt	cubic metre/m^3	4.732×10^{-4}
	UK gallon/gal	cubic metre/m^3	4.546×10^{-3}
	US gallon/gal	cubic metre/m^3	3.785×10^{-3}
Angle	degree/°	radian/rad	1.745×10^{-2}
Mass	tonne (UK)	kilogram/kg	1×10^3
	ton	kilogram/kg	1.016×10^3
	hundredweight/cwt	kilogram/kg	5.080×10^1
	stone	kilogram/kg	6.350
	pound/lb	kilogram/kg	0.454
	ounce/oz	kilogram/kg	2.835×10^{-2}
Energy	erg	joule/J	1×10^{-7}
	kilowatt hour/kWh	joule/J	3.6×10^6
Pressure	bar/b	pascal/Pa	1×10^5
	mm Hg	pascal/Pa	1.332×10^2
Radioactivity	curie/Ci	becquerel/Bq	3.7×10^{10}
Temperature	centigrade/°C	kelvin/K	$C + 273.15$
	Fahrenheit/°F	kelvin/K	$^5/_9\,(F + 459.7)$

[a]Metric units are given in italics. To get from a measurement in the old unit to a measurement in the SI unit, multiply by the conversion factor.

hydrogen atoms. In other words, the mass of 1 mole of a substance is its molecular mass in grams.

When working out concentrations of solutions it is probably best to stick to these units, since most glassware is still calibrated in litres and small balances in grams.

The molarity M of a solution is obtained as follows:

$$M = \frac{\text{Number of moles}}{\text{Solution volume (l)}}$$

$$= \frac{\text{Mass (g)}}{\text{Molecular mass} \times \text{Solution volume (l)}}$$

Example 10.2

A solution contains 23 g of copper sulphate ($CuSO_4$) in 2.5 l of water. What is its concentration?

Solution

$$\begin{aligned}\text{Concentration} &= 23/((63.5 + 32 + 64) \times 2.5) \\ &= 5.768 \times 10^{-2}\,M \\ &= 5.8 \times 10^{-2}\,M \text{ (two significant figures)}\end{aligned}$$

Non-metric units

Non-metric units, such as those based on the old Imperial scale, are also given in Table 10.4. Again you must simply multiply your measurement by the conversion factor. However, they are more difficult to convert to SI, since they must be multiplied by factors which are not just powers of 10. For instance,

$$6 \text{ ft} = 6 \times 3.048 \times 10^{-1} \text{ m} = 1.83 \text{ m}$$

Note that the answer was given as 1.83 m, not the calculated figure of 1.8288 m. This is because a measure of 6 ft implies that the length was measured to the nearest inch. The answer we produced is accurate to the nearest centimetre, which is the closest SI unit.

If you have to convert square or cubic measures into metric, simply multiply by the conversion factor to the power of 2 or 3. So 12 $ft^3 = 12 \times (3.038 \times 10^{-1})^3\ m^3 = 3.4 \times 10^{-1}\ m^3$ to two significant figures.

10.4 Combining values

Once measurements have been converted into SI units with exponents, they are extremely straightforward to combine using either pencil and paper or calculator (most calculators use exponents nowadays). When multiplying two measurements, for instance, you simply multiply the initial numbers, add the exponents and multiply the units together. If the multiple of the two initial numbers is greater than 10 or less than 1, you simply add or subtract 1 from the exponent. For instance,

$$\begin{aligned}2.3 \times 10^2\,\text{m} \times 1.6 \times 10^3\,\text{m} &= (2.3 \times 1.6) \times 10^{(2-3)}\,\text{m}^2 \\ &= 3.7 \times 10^5\,\text{m}^2\end{aligned}$$

Notice that the area is given to two significant figures because that was the degree of precision with which the lengths were measured. Similarly,

$$\begin{aligned}2.3 \times 10^2\,\text{m} \times 6.3 \times 10^{-4}\,\text{m} &= (2.3 \times 6.3) \times 10^{(2-4)}\,\text{m}^2 \\ &= 1.4 \times 10^1 \times 10^{-2}\,\text{m}^2 \\ &= 1.4 \times 10^{-1}\,\text{m}^2\end{aligned}$$

In the same way, when dividing one measurement by another you divide the first initial number by the second, subtract the second exponent from the first, and divide the first unit by the second.

$$\text{Therefore } (4.8 \times 10^3\ \text{m})/(1.5 \times 10^2\ \text{s}) = (4.8/1.5) \times 10^{(3-2)}\ \text{ms}^{-1}$$
$$= 3.2 \times 10^1\ \text{ms}^{-1}$$

10.5 Expressing the answer

When you have completed all calculations, you must be careful how you express your answer. First, it should be given to the same level of precision as the *least* accurate of the measurements from which it was calculated. This book and many statistical packages use the following convention: the digits 1 to 4 go down, 6 to 9 go up and 5 goes to the nearest even digit. Here are some examples:

0.343	becomes	0.34	to	2	significant figures
0.2251	becomes	0.22	to	2	significant figures
0.6354	becomes	0.64	to	2	significant figures

Second, it is sometimes a good idea to express it using a prefix. So if we work out from figures given to two significant figures that a pressure is 2.678×10^6 Pa, it should be expressed as 2.7 MPa. Always adjust the degree of precision *at the end* of the calculation.

10.6 Doing all three steps

The various steps can now be carried out to manipulate data to reliably derive further information. It is important to carry out each step in its turn, producing an answer before going on to the next step in the calculation. Doing all the calculations at once can cause confusion and lead to silly mistakes.

Example 10.3

A sample of heartwood taken from an oak tree was 12.1 mm long by 8.2 mm wide by 9.5 mm deep and had a wet mass of 0.653 g. What was its density?

Solution

Density has units of mass (in kg) per unit volume (in m^3). Therefore the first thing to do is to convert the units into kg and m. The next thing to do is to calculate the volume in m^3. Only then can the final calculation be performed. This slow building up of the calculation is ponderous but is the best way to avoid making mistakes.

$$\text{Mass} = 6.53 \times 10^{-4}\ \text{kg}$$
$$\text{Volume} = 1.21 \times 10^{-2} \times 8.2 \times 10^{-3} \times 9.5 \times 10^{-3}$$
$$= 9.4259 \times 10^{-7}\ \text{m}^3$$

$$\text{Density} = \frac{\text{Mass}}{\text{Volume}} = \frac{6.53 \times 10^{-4}}{9.4259 \times 10^{-7}}$$
$$= 0.6928 \times 10^3 \text{ kg m}^{-3}$$
$$= 6.9 \times 10^2 \text{ kg m}^{-3}$$

Notice that the answer is given to two significant figures, like the dimensions of the sample.

10.7 Constants and formulae

Frequently, raw data on their own are not enough to work out other important quantities. You may need to include physical or chemical constants in your calculations, or insert your data into basic mathematical formulae. Table 10.5

Table 10.5 Some useful constants and formulae

Physical constants	
Density of water	$= 1000 \text{ kg m}^{-3}$
Density of air	$= 1.2 \text{ kg m}^{-3}$
Specific heat of water	$= 4.2 \times 10^3 \text{ J K}^{-1} \text{ kg}^{-1}$
Chemical constants	
1 mol	$= 6 \times 10^{23}$ particles
Mass of 1 mol	= molecular mass (g) = 10^{-3} × molecular mass (kg)
Volume of 1 mol of gas	$= 24 \text{ l} = 2.4 \times 10^{-2} \text{ m}^3$ (at room temperature and pressure)
1 molar solution (1 *M*)	$= 1 \text{ mol l}^{-1} = 1000 \text{ mol m}^{-3}$
1 normal solution (1 *N*)	$= 1 \text{ mol l}^{-1} = 1000 \text{ mol m}^{-3}$ of ions
	$\text{pH} = -\log_{10}[H^+]$
Composition of air	= 78.1% nitrogen, 20.9% oxygen, 0.93% argon and 0.03% carbon dioxide, plus traces of others, by volume
Mathematical formulae	
Area of a circle of radius R	$= \pi R^2$
Volume of a sphere of radius R	$= {}^4/_3 \pi R^3$
Area of a sphere of radius R	$= 4\pi R^2$
Volume of a cylinder of radius R and height H	$= \pi R^2 H$
Volume of a cone of radius R and height H	$= {}^1/_3 \pi R^2 H$
Mathematical constants	
$\pi = 3.1416$	
$\log_e x = 2.30 \log_{10} x$	

is a list of some useful constants and formulae. Many of them are worth memorising.

Example 2.4

A total of 25 micropropagated plants were grown in a 10 cm diameter Petri dish. At what density were they growing?

Solution

The first thing to calculate is the area of the Petri dish. Since its diameter is 10 cm, its radius R will be 5 cm (or 5×10^{-2} m). A circle's area A is given by the formula $A = \pi R^2$. Therefore

$$\begin{aligned} \text{Area} &= 3.1416 \times (5 \times 10^{-2})^2 \\ &= 7.854 \times 10^{-3}\ \text{m}^2 \end{aligned}$$

The density is the number per unit area, so

$$\begin{aligned} \text{Density} &= 25/(7.854 \times 10^{-3}) \\ &= 3.183 \times 10^{3}\ \text{m}^{-2} \\ &= 3.2 \times 10^{3}\ \text{m}^{-2}\ \text{(two significant figures)} \end{aligned}$$

10.8 Using calculations

Once you can reliably perform calculations, you can use them for far more than just working out the results of your experiments from your raw data. You can use them to put your results into perspective or extrapolate from your results into a wider context. You can also use calculations to help design your experiments: to work out how much of each ingredient you need, or how much the experiment will cost. But even more usefully, they can help you to work out whether a particular experiment is worth attempting in the first place. Calculations are thus an invaluable tool for the research biologist to help save time, money and effort. They don't even have to be very exact calculations. Often, all that is required is to work out a rough or ballpark figure.

Example 10.5

Elephants are the most practical form of transport through the Indian rainforest because of the rough terrain; the only disadvantage is their great weight. A scientific expedition needs to cross a bridge with a weight limit of 10 tonnes, in order to enter a nature reserve. Will the elephants be able to cross this bridge safely?

Solution

You are unlikely, in the rainforest, to be able to look up or measure the weight of an elephant, but most people have some idea of just how big they are. Since the mass of an object is equal to volume × density, the first thing to calculate is the volume. What is the volume of an elephant? Well, elephants are around 2–3 m long and have a (very roughly) cylindrical body of diameter, say, 1.5 m (so the radius = 0.75 m). The volume of a cylinder is given by $V = \pi R^2 L$, so with these figures the volume of the elephant is approximately

$$V = \pi \times 0.75^2 \times 2 \quad \text{up to} \quad \pi \times 0.75^2 \times 3$$
$$V = 3.53 - 5.30 \text{ m}^3$$

The volume of the legs, trunk, etc. is very much less and can be ignored in this rough calculation. So what is the density of an elephant? Well, elephants (like us) can just float in water and certainly swim, so they must have about the same density as water, 1000 kg m^{-3}. The approximate mass of the elephant is therefore

$$\text{Mass} = 1000 \times (3.5 \text{ to } 5.3)$$
$$= 3530 - 5300 \text{ kg}$$

Note, however, that the length of the beast was estimated to only one significant figure, so the weight should also be estimated to this low degree of accuracy. The weight of the elephant will be (4–5) × 10^3 kg or 4–5 tonnes. (Textbook figures for weights of elephants range from 3 to 7 tonnes.) The bridge should easily be able to withstand the weight of an elephant.

This calculation would not have been accurate enough to determine whether our elephant could cross a bridge with weight limit 4.5 tonnes. It would have been necessary to devise a method of weighing it.

10.9 Logarithms, graphs and pH

10.9.1 Logarithms to base 10

Though scientific notation (such as 2.3×10^4) is a good way of expressing large and small numbers (such as 23 000), it is a bit clumsy since the numbers consist of two parts, the initial number and the exponent. Nor does it help very large and very small numbers to be conveniently represented on the same graph. For instance, if you plot the relationship between the numbers of bird species in woods of areas 100, 1000, 10 000, 100 000 and 1 000 000 m^2 (Figure 10.1a), most of the points will be congested at the left.

These problems can be overcome by the use of **logarithms**. Any number can be expressed as a single **exponent**, as 10 to the power of a second number, e.g. 23 000 = $10^{4.362}$ The 'second number' (here 4.362) is called the logarithm to base 10 ($\log_{10}$) of the first, so that

exponent
A power of 10 which allows large or small numbers to be readily expressed and manipulated.

$$4.362 = \log_{10} 23\,000$$

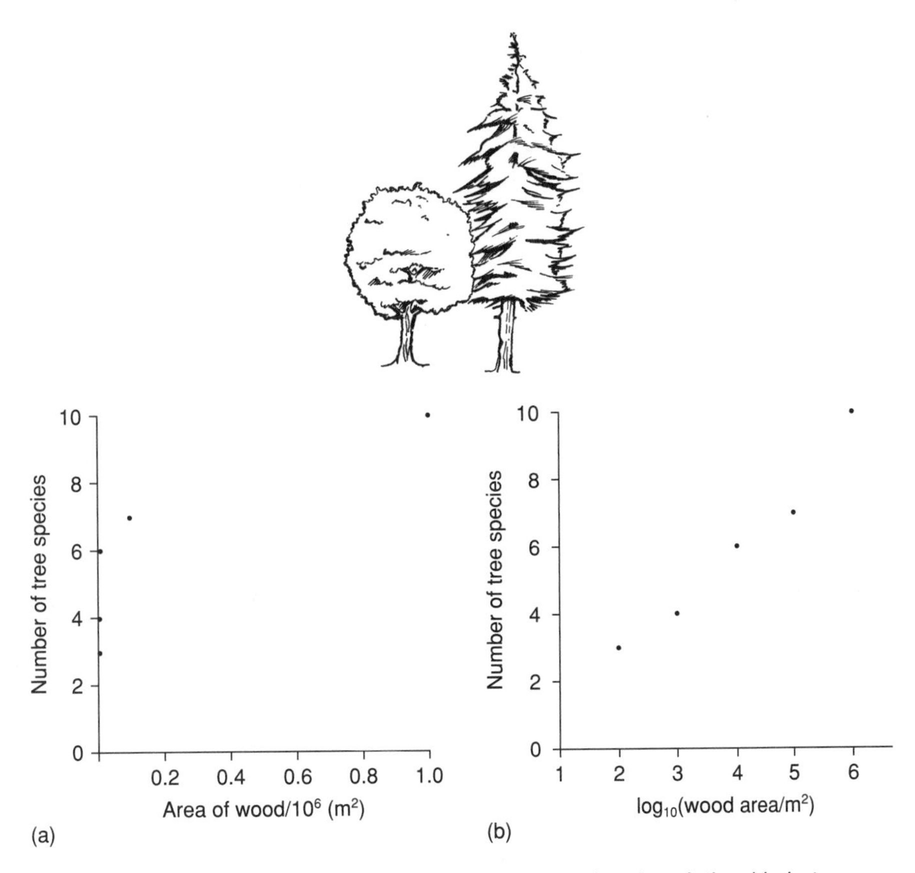

Figure 10.1 Using logarithms. (a) A simple graph showing the relationship between the size of woods and the number of tree species they contain; the points are hopelessly congested at the left of the plot. **(b)** Plotting number of species against $\log_{10}$ (area) spreads the data out more evenly.

Numbers above 1 have a positive logarithm, whereas numbers below 1 have a negative logarithm, e.g.

$$0.0045 = 10^{-2.347} \text{ so } -2.347 = \log_{10} 0.0045$$

Logarithms to the base 10 of any number can be found simply by pressing the log button on your calculator, and can be converted back to real numbers by pressing the 10^x button on your calculator.

Properties and uses of logarithms

The most important property of logarithms is that if numbers have a constant ratio between them, their logarithms will differ by a constant amount. Hence the numbers 1, 10 and 100, which differ by ratios of 10, have logarithms of 0, 1 and 2, which differ by 1 each time. This gives them some useful mathematical properties, which can help us work out relationships between variables, as we saw in Chapter 6. However, it also gives them two immediate uses.

Use of logarithms for graphs

Logarithms allow very different quantities to be compared and plotted on the same graph. For instance, you can show the relationship between wood area and number of tree species (Figure 10.1a) more clearly by plotting species number against $\log_{10}$ (area) (Figure 10.1b).

pH

The single most important use of logarithms in biology is in the units for acidity. The unit pH is given by the formula

$$\text{pH} = -\log_{10}[\text{H}^+] \qquad (10.1)$$

where $[\text{H}^+]$ is the hydrogen ion concentration in moles per litre (*M*). Therefore, a solution containing 2×10^{-5} mole (mol) of hydrogen ions per litre will have a pH of $-\log_{10}(2 \times 10^{-5}) = 4.7$.

Example 10.6 A solution has a pH of 3.2. What is the hydrogen ion concentration?

Solution

The hydrogen ion concentration is $10^{-3.2} = 6.3 \times 10^{-4}$ *M*.

10.9.2 Natural logarithms

natural logarithm ($\log_e$ or ln)
A function of a variable y such that if $y = e^x$ then $x = \log_e y$ or $\ln y$.

Logarithms can be calculated for other bases as well as 10. Other important types of logarithms are **natural logarithms** ($\log_e$ or ln) in which numbers that differ by the ratio 2.718 (which is given the letter e) have logs that differ by 1. Thus ln 2.718 $\times$ 1. As we saw in Chapter 5, natural logarithms are particularly useful when describing and investigating exponential increases in populations or exponential decay in radioactivity.

To convert from a number to its natural logarithm, you should press the ln button on your calculator. To convert back, you should press the e^x button.

10.10 Self-assessment problems

Problem 10.1

What are the SI units for the following measurements?

(a) Area
(b) The rate of height growth for a plant
(c) The concentration of red cells in blood
(d) The ratio of the concentrations of white and red cells in blood

Problem 10.2

How would you express the following quantities using appropriate prefixes?

(a) 192 000 000 N
(b) 0.000 000 102 kg
(c) 0.000 12 s
(d) 21.3 cm

Problem 10.3

How would you express the following quantities in scientific notation using appropriate exponents?

(a) 0.000 046 1 J
(b) 461 000 000 s

Problem 10.4

How would you express the following quantities in scientific notation using the appropriate exponents?

(a) 3.81 GPa
(b) 4.53 mW
(c) 364 mJ
(d) 4.8 mg
(e) 0.21 pg

Problem 10.5

Convert the following to SI units expressed in scientific notation.

(a) 250 tonnes
(b) 0.3 bar
(c) 24 angstroms

Problem 10.6

Convert the following into SI units.

(a) 35 yards
(b) 3 feet 3 inches
(c) 9.5 square yards

Problem 10.7

Perform the following calculations.

(a) 1.23×10^3 m $\times$ 2.456×10^5 m
(b) (2.1×10^{-2} J) / (4.5×10^{-4} kg)

Problem 10.8

Give the following expressions in prefix form and to the correct degree of precision.

(a) 1.28×10^{-3} mol to two significant figures
(b) 3.649×10^{8} J to three significant figures
(c) 2.423×10^{-7} m to two significant figures

Problem 10.9

A blood cell count was performed. Within the box on the slide, which had sides of length 1 mm and depth of 100 μm, there were 652 red blood cells. What was the concentration of cells (in m^{-3}) in the blood?

Problem 10.10

An old-fashioned rain gauge showed that 0.6 in. of rain had fallen on an experimental plot of area of 2.6 ha. What volume of water had fallen on the area?

Problem 10.11

What is the concentration of a solution of 25 g of glucose (formula $C_6H_{12}O_6$) in a volume of 2000 ml of water?

Problem 10.12

An experiment to investigate the basal metabolic rate of human beings showed that in 5 minutes the subject breathed out 45 l of air into a Douglas bag. The oxygen concentration in this air had fallen from 19.6% by volume to 16.0%, so it contained 3.6% CO_2 by volume. What was the mass of this CO_2 and at what rate had it been produced?

Problem 10.13

A chemical reaction heated 0.53 l of water at 2.4 K. How much energy had it produced?

Problem 10.14

An experiment which must be repeated around 8 times requires 80 ml of a 3×10^{-3} M solution of the substance X. Given that X has a molecular mass of 258 and costs £56 per gram, and given that your budget for the year is £2000, do you think you will be able to afford to do the experiment?

Problem 10.15

It has been postulated that raised bogs may be major producers of methane and, because methane is a greenhouse gas, therefore an important cause of the greenhouse effect. A small microcosm experiment was carried out to investigate the rate at which methane is produced by a raised bog in North Wales. This showed that the rate of production was 21 ml m^{-2} per day. Given that (1) world production of CO_2 by burning fossil fuels is 25 Gt per year, (2) weight for weight, methane is said to be three times more efficient a greenhouse gas than CO_2 and (3) there is 3.4×10^{6} km^2 of blanket bog in the world, what do you think of this idea?

Problem 10.16

Calculate $\log_{10}$ of

(a) 45
(b) 450
(c) 0.000 45
(d) 1 000 000
(e) 1

Problem 10.17

Reconvert the following logarithms to base 10 back to numbers.

(a) 1.4
(b) 2.4
(c) −3.4
(d) 4
(e) 0

Problem 10.18

Calculate the pH of the following solutions.

(a) 3×10^{-4} M HCl
(b) 4×10^{-6} M H_2SO_4

Problem 10.19

Calculate the mass of sulphuric acid (H_2SO_4) in 160 ml of a solution which has a pH of 2.1.

Problem 10.20

Calculate the natural logarithm of

(a) 30
(b) 0.024
(c) 1

Problem 10.21

Convert the following natural logarithms back to numbers.

(a) 3
(b) −3
(c) 0

Glossary

ANCOVA Abbreviation for analysis of covariance: a series of tests that determine whether there are significant differences between groups, while accounting for variability in other measurements.

ANOVA Abbreviation for analysis of variance: a widely used series of tests which can determine whether there are significant differences between groups.

association A numerical link between two sets of measurements.

binomial distribution The pattern by which the sample frequencies in two groups tends to vary.

blocking A method of eliminating confounding variables by spreading replicates of the different samples evenly among different blocks.

category A character state which cannot meaningfully be represented by a number.

causal relationship Relationship between two variables whereby one affects the other but is not itself affected.

chi-squared (χ^2) A statistical test which determines whether there are differences between real and expected frequencies in one set of categories, or associations between two sets of categories.

confidence limits Limits between which estimated parameters have a defined likelihood of occurring. It is common to calculate 95% confidence limits, but 99% and 99.9% confidence limits are also used. The range of values between the upper and lower limits is called the confidence interval.

confounding variables Variables which invalidate an experiment if they are not taken into account.

contingency table A table showing the frequencies of two sets of character states, which allows you to calculate expected values in a chi-squared test for association.

correlation A statistical test which determines whether there is linear association between two sets of measurements.

critical values Tabulated values of test statistics; usually if the absolute value of a calculated test statistic is greater than or equal to the appropriate critical value, the null hypothesis must be rejected.

data Observations or measurements you have taken (your results) which are used to work things out about the world.

degrees of freedom (DF) A concept used in parametric statistics, based on the amount of information you have when you examine samples. The number of degrees of freedom is generally the total number of observations you make minus the number of parameters you estimate from the samples.

dependent variable A variable in a regression which is affected by another variable.

descriptive statistics Statistics which summarise the distribution of a single set of measurements.

distribution The pattern by which a measurement or frequency varies.

error bars Bars drawn upwards and downwards from the mean values on graphs; error bars can represent the standard deviation or the standard error.

estimate A parameter of a population which has been calculated from the results of a sample.

exponent A power of 10 which allows large or small numbers to be readily expressed and manipulated.

exponential relationship A relationship which follows the general equation $y = ae^{bx}$. If $b > 0$ this is exponential growth; if $b < 0$ this is exponential decay.

frequency The number of times a particular character state turns up.

general linear model (GLM) A series of tests which combine ANOVA and regression analysis to allow powerful analysis of complex data sets.

Imperial Obsolete system of units from the United Kingdom.

independent variable A variable in a regression which affects another variable but is not itself affected.

interaction A synergistic or inhibitory effect between two factors which can be picked up by using two-way ANOVA.

intercept The point where a straight line crosses the y-axis.

logarithm to base 10 ($\log_{10}$) A function of a variable y such that if $y = 10^x$ then $x = \log_{10} y$.

logistic regression A statistical test which analyses how a binary outcome is affected by other numerical characteristics.

mean (μ) The average of a population. The estimate of μ is called $\bar{x}$.

mean square The variance due to a particular factor in analysis of variance (ANOVA).

measurement A character state which can meaningfully be represented by a number.

median The central value of a distribution (or average of the middle points if the sample size is even).

metric Units based on the metre, second and kilogram but not necessarily SI.

natural logarithm ($\log_e$ or ln) A function of a variable y such that if $y = e^x$ then $x = \log_e y$ or $\ln y$.

non-parametric test A statistical test which does not assume that data is normally distributed, but instead uses the ranks of the observations.

normal distribution The usual symmetrical and bell-shaped distribution pattern for measurements that are influenced by large numbers of factors.

null hypothesis A preliminary assumption in a statistical test that the data shows no differences or associations. A statistical test then works out the probability of obtaining data similar to your own by chance.

parameter A measure, such as the mean and standard deviation, which describes or characterises a population. These are usually represented by Greek letters.

parametric test A statistical test which assumes that data are normally distributed.

placebo A non-active control treatment used in drug trials.

population A potentially infinite group on which measurements could be taken. Parameters of populations usually have to be estimated from the results of samples.

post hoc tests Statistical tests carried out if an analysis of variance is significant; they are used to determine which groups are different from each other.

power relationship A relationship which follows the general equation $y = ax^b$.

prefix A multiple or divisor of 1000 which allows large or small numbers to be readily expressed.

quartiles Upper and lower quartiles are values exceeded by 25% and 75% of the data points, respectively.

rank Numerical order of a data point.

regression A statistical test which analyses how one set of measurements is (usually linearly) affected by another.

replicates The individual data points.

replication The use of large numbers of measurements to allow one to estimate population parameters.

sample A subset of a possible population on which measurements are taken. These can be used to estimate parameters of the population.

scatter plot A point graph between two variables which allows one to visually determine whether they are associated.

scientific notation A method of representing large or small numbers, giving them as a number between 1 and 10 multiplied by a power of 10.

SI Système International: the common standard of units used in modern science based on the metre, second and kilogram.

significance probability The chances that a certain set of results could be obtained if the null hypothesis were true.

significant difference A difference which has less than a 5% probability of having happened by chance.

skewed data Data with an asymmetric distribution.

slope The gradient of a straight line.

standard deviation (σ) A measure of spread of a group of measurements: the amount by which on average they differ from the mean. The estimate of σ is called s.

standard error (SE) A measure of the spread of sample means: the amount by which they differ from the true mean. Standard error equals standard deviation divided by the square root of the number in the sample. The estimate of SE is called i.

standard error of the difference (i_d) A measure of the spread of the difference between two estimated means.

statistic An estimate of a population parameter, found by random sampling. Statistics are represented by Latin letters.

***t* distribution** The pattern by which sample means of a normally distributed population tend to vary.

***t* tests** Statistical tests which analyse whether there are differences between measurements on a single population and an expected value, between paired measurements, or between two unpaired sets of measurements.

transformation A mathematical function used to make the distribution of data more symmetrical and so make parametric tests valid.

two-tailed tests Tests which ask merely whether observed values are different from an expected value or each other, not whether they are larger or smaller.

type 1 error The detection of an apparently significant difference or association, when in reality there is no difference or association between the populations.

type 2 error The failure to detect a significant difference or assocation, when in reality there is a difference or association between the populations.

variance A measure of the variability of data: the square of their standard deviation.

Further reading

Statistics is a huge field, so this short book has of necessity been superficial and selective in the material it covers. For more background and for more information about the theory of statistics the reader is referred to the following titles. The books vary greatly in their approach and in the degree of mathematical competence required to read them, so you should choose your texts carefully. Sokal and Rohlf (2009) is really the bible of biological statistics, but most students will struggle with its high mathematical tone. Field (2009) is much easier to understand. It is based on SPSS and, though really for psychologists, gives a lot of useful background to the more complex GLM and data exploration techniques. Heath (1995) and Ruxton and Colegrave (2003) gives useful advice on experimental design. In contrast, if you are struggling even with the complexity of this book, a very simple introduction to statistical thinking (without any equations!) is Rowntree (1991).

Field, A. (2009) *Discovering Statistics Using SPSS*. Sage Publications, London.

Heath, D. (1995) *An Introduction to Experimental Design and Statistics for Biology*. UCL Press, London.

Rowntree, D. (1991) *Statistics without Tears*. Penguin Books, London.

Ruxton, G. D. and Colegrave, N. (2003) *Experimental Design for the Life Sciences*. Oxford University Press, Oxford.

Sokal, R. R. and Rohlf, F. J. (2009) *Introduction to Biostatistics, second edition*. Dover Publications Inc., San Francisco.

Solutions

Chapter 2

Problem 2.1

95% will have heart rates around 75 ± (1.96 × 11), i.e. between 53 and 97 beats per minute.

Problem 2.2

Mean = 5.71 g, $s = 0.33$ g.

Problem 2.3

(a) Mean = 5.89, $s = 0.31$, $\overline{SE} = 0.103$, 95% CI = 5.89 ± (2.306 × 0.103) = 5.65 to 6.13.
(b) Mean = 5.95, $s = 0.45$, $\overline{SE} = 0.225$, 95% CI = 5.95 ± (3.182 × 0.225) = 5.23 to 6.67. The 95% confidence interval is three times wider than (a).

Problem 2.4

(a) Mean (s) = 3.00 (0.47) kg, $n = 25$, $\overline{SE} = 0.093$ kg.
(b) The bar chart is shown in Figure A1.

Chapter 3

Problem 3.1

Since this is proportional data, with investment into roots given as a proportion of total mass, it must be subjected to an **arcsin** $\sqrt{x}$ transformation.

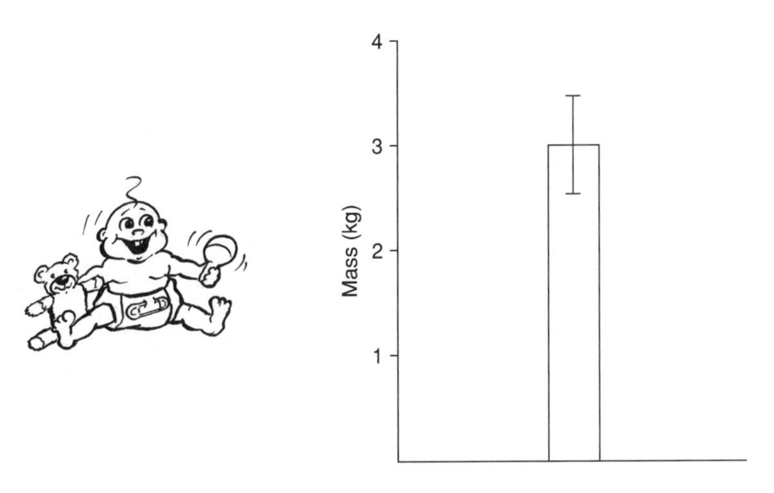

Figure A1 Mean birthweight. Error bar represents standard deviation.

Problem 3.2

Exploring the data in SPSS shows it to be strongly positively skewed with many more small species than large ones.

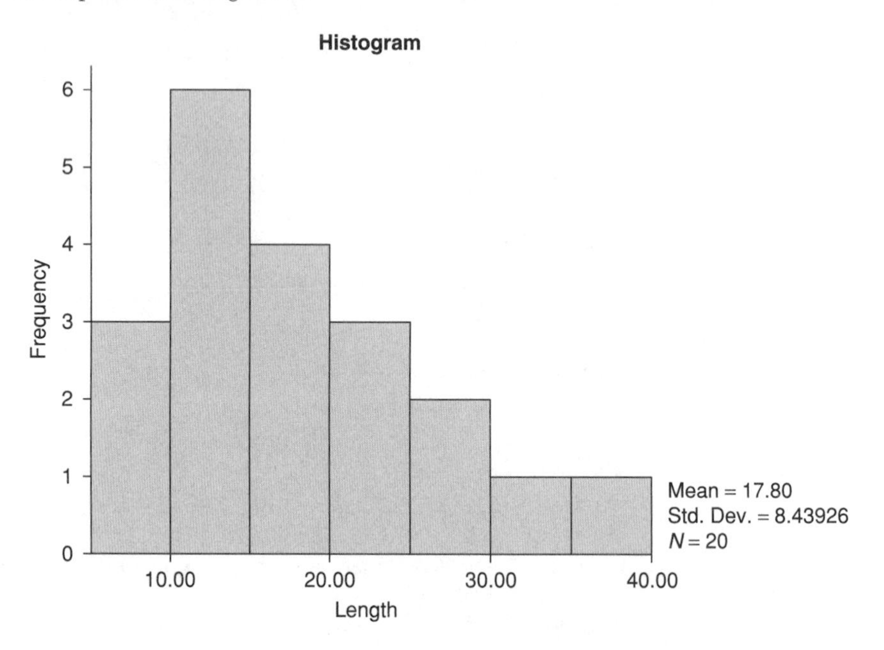

To obtain a more symmetrical distribution you should carry out a log transformation to give results like those shown below.

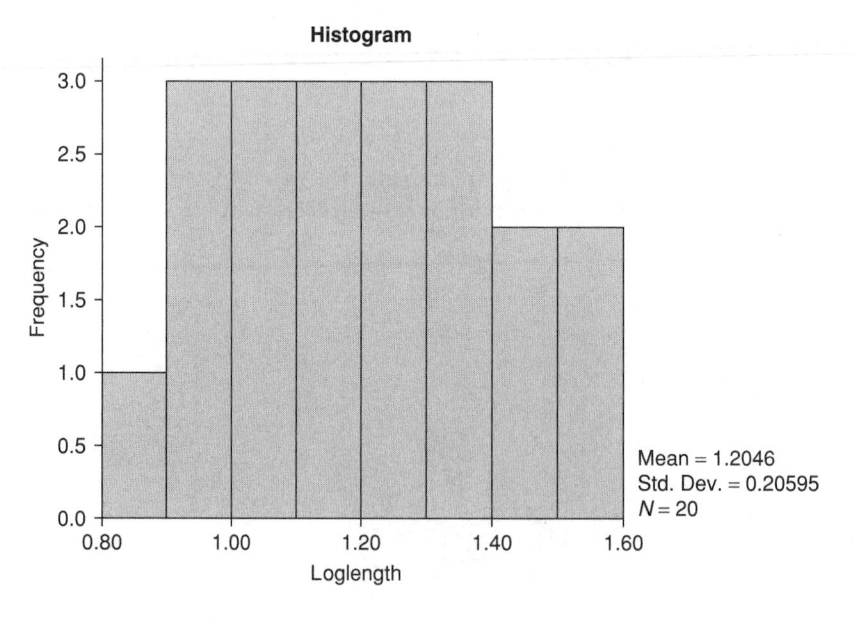

Note that Kolgomorov–Smirnov tests show that neither distribution is significantly different from normal, demonstrating that it is important to actually look at the distribution of data.

Chapter 4

Problem 4.1

The null hypothesis is that the mean score = 58%. The mean score of the students is $\bar{x} = 58.36$ with $s = 13.70$ and $\overline{SE} = 2.74$. The score seems higher than expected but in the one-sample t test, $t = (58.36 - 58)/2.74 = 0.13$ to two decimal places. The absolute value of t, 0.13, is therefore well below the value of 2.064 needed for significance at 24 degrees of freedom. The SPSS and MINITAB give a significance probability of 0.897. This is much greater than the value of 0.05 needed for significance.

Therefore students did not perform significantly differently from expected.

Problem 4.2

The null hypothesis is that the mean mass of tomatoes is 50 g. Looking at descriptive statistics, $\bar{x} = 44.1$ g, $s = 8.6$ g, $\overline{SE} = 2.15$ kg. This seems well below 50 g. In the one-sample t test to determine whether the mean mass is significantly different from 50 g, $t = (44.1 - 50)/2.15 = -2.74$ to two decimal places. The absolute value of t, 2.74, is well above the value of 2.131 required for significance at 15 degrees of freedom (remember that is the magnitude of t and not whether it is positive or negative that matters).

SPSS and MINITAB give a significance probability 0.016, well below the 0.05 needed for significance.

Therefore the tomatoes are significantly lighter than the expected 50 g. The 95% confidence interval for mass is $44.1 \pm (2.131 \times 2.15) = 39.5$ to 48.7 g.

Problem 4.3

The null hypothesis is that students' scores after the course were the same as before it. The mean score was 58.1 before and 53.8 after. The scores seem to be worse afterwards, but to find whether the difference is significant you need to carry out a paired t test. This shows a mean difference $d = -4.3$, $s = 5.7$ and $\overline{SE}_d = 1.79$. Therefore $t = -4.3/1.79 = -2.40$ to two decimal places. Its absolute value, 2.40, is larger than the value of 2.306 needed for significance at $9 - 1 = 8$ degrees of freedom.

SPSS and MINITAB give the significance probability as 0.040, below the 0.05 needed for significance.

Therefore the course did have a significant effect. After the course most students got worse marks!

The 95% confidence interval for the difference is $-4.3 \pm (2.306 \times 1.79) = -0.2$ to -8.4.

Problem 4.4

(a) The null hypothesis is that pH was the same at dawn and dusk. The mean at dawn was 5.54 (with $s = 0.71$ and $\overline{SE}_d = 0.203$) and at dusk was 6.45 (with $s = 0.64$ and $\overline{SE}_d = 0.193$). The pH seems to be higher at dusk but to find out you need to carry out a two sample t test. Using equations 3.5 and 3.6 we can calculate that

$$t = (5.54-6.45)/\sqrt{((0.204)^2 + (0.193)^2)} = -0.91/0.282 = -3.20$$

to two decimal places. Its absolute value, 3.20, is larger than the value of 2.080 needed for significance at $12 + 11 - 2 = 21$ degrees of freedom.

MINITAB and SPSS give the significance probability as 0.004, well below the 0.05 needed for significance.

Therefore the pH is significantly different at dawn and dusk, being significantly higher at the end of the day.

(b) You cannot use a paired t test, because the cells which you measured are not identifiably the same. Indeed, different numbers of cells were examined at dawn and dusk.

Problem 4.5

The null hypothesis is that the control and supported plants have the same mean yield. The mean yield of controls was 10.28 (with $s = 1.60$ and $\overline{SE}_d = 0.36$) and supported plants was 10.06 (with $s = 1.55$ and $\overline{SE}_d = 0.35$). The yield seems to be higher in the controls but to find if this is a significant difference you need to carry out a two-sample t test. Using equations 3.5 and 3.6 we can calculate that

$$t = (10.28-10.06)/\sqrt{((0.36)^2 + (0.35)^2)} = 0.22/0.282 = 0.43$$

to two decimal places. Its absolute value, 0.43, is smaller than the value of 2.025 needed for significance at $20 + 20 - 2 = 38$ degrees of freedom.

SPSS and MINITAB give the significance probability as 0.669, well above the 0.05 needed for significance.

Therefore the yield is not significantly different between the control and supported plants.

Problem 4.6

(a) The data cannot be transformed to be normally distributed because the weight of the deer at the start is bimodally distributed as can be seen in the SPSS plot below; the males are on average much heavier than the females.

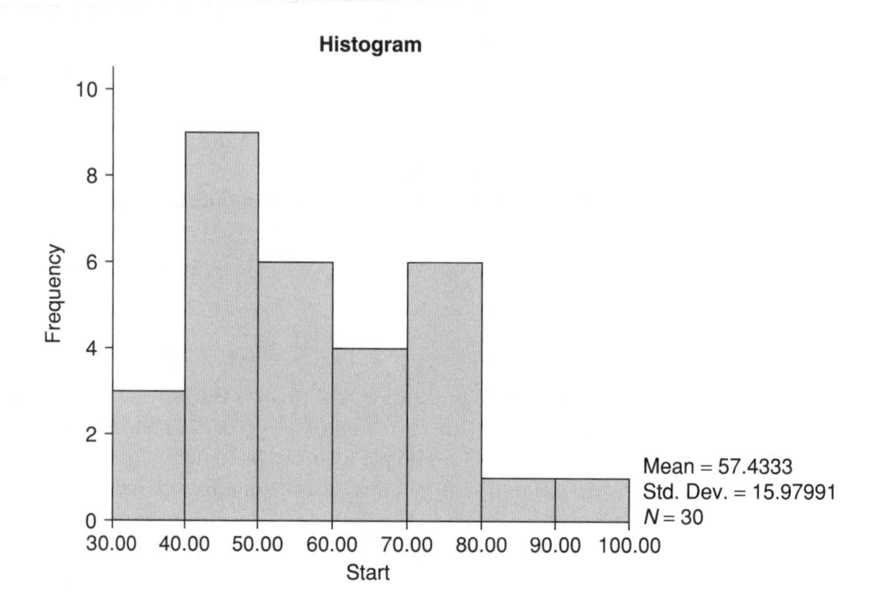

(b) The null hypothesis is that the weight of the deer is the same at the end as at the start of the summer. Descriptive statistics given by SPSS or MINITAB show that the median weight at the end is actually higher (57.0 kg) than at the start (53.5 kg) of summer, but is this a significant difference? Carrying out the Wilcoxon matched pairs test shows that the sum of negative ranks (63) is far less than that of positive ranks (402). Looking up in Table S5 for the Wilcoxon T distribution, shows that the lower value, 63, is far less than the value of 137 needed for significance for 30 matched pairs of data. SPSS and MINITAB give the significance probability as 0.000, well below the 0.05 needed for significance.

Therefore the deer did weigh significantly different at the end of the summer compared with the start; they were heavier.

Problem 4.7

The null hypothesis is that the animal spent equal time pacing when in the two cages. The sum of the ranks is lower for cage 2 (132.5) than for cage 1 (167.5), and results in a value for U of 54.5, but does this show a significant difference? Looking up in Table S4 shows that for $n_1 = 12$ and $n_2 = 12$ the critical value of U is 107, so our value is above that needed for significance. SPSS and MINITAB give the significance probability as 0.294, well above the value of 0.05 needed for significance.

Therefore there is no significant difference in the time the animal spends pacing in the two cages.

Problem 4.8

(a) You should use the Mann–Whitney U test because there are two groups of students, those given the drug and those the placebo, and there was no matching into pairs of these two groups.

(b) The null hypothesis is that there is no difference in the scarring between patients given the drug and those that had been given the placebo. The mean ranks of the two groups was different (514 and 306) but is this difference significant? A Mann–Whitney U test gives a value of U of 96. Looking up in Table S4 shows that for $n_1 = 20$ and $n_2 = 20$ the critical value of U is 273, so our value is well below that needed for significance. SPSS and MINITAB give the significance probability as 0.004, well below the value of 0.05 needed for significance.

Therefore there is a significant difference in the level of scarring between patients given the drug and those given the placebo; those given the drug actually had less scarring.

Chapter 5

Problem 5.1

The null hypothesis is that the mean activity is the same at each point in time. To test whether this is the case you should carry out one-way ANOVA as well as determine the descriptive statistics for the five time points.

For example SPSS produces the following results.

Descriptives

Activity

					95% Confidence interval for mean			
	N	Mean	Std. deviation	Std. error	Lower bound	Upper bound	Minimum	Maximum
0.00	6	2.9500	0.53572	0.21871	2.3878	3.5122	2.20	3.80
1.00	6	3.1667	0.48442	0.19777	2.6583	3.6750	2.70	4.00
2.00	6	3.9000	0.44272	0.18074	3.4354	4.3646	3.50	4.70
4.00	6	4.5167	0.43089	0.17591	4.0645	4.9689	3.90	5.10
8.00	6	3.2500	0.32711	0.13354	2.9067	3.5933	2.70	3.60
Total	30	3.5567	0.71856	0.13119	3.2884	3.8250	2.20	5.10

ANOVA

Activity

	Sum of squares	df	Mean square	F	Sig.
Between groups	9.922	4	2.481	12.276	0.000
within groups	5.052	25	0.202		
total	14.974	29			

It is clear that the mean activity rises from 2.95 at the start to a peak of 4.52 after 4 hours, but are these changes significant? Looking at the ANOVA table, SPSS gives a high value of $F = 12.276$, while Sig. = 0.000, well below the 0.05 level needed for significance. Therefore activity is different at different times. But at which times is the activity raised from the control. To find out you must perform a **Dunnett** test in SPSS which will produce the following results.

Multiple comparisons

Dependent variable: activity
Dunnett t (2-sided)[a]

					95% Confidence interval	
(I) hours	(J) hours	Mean difference ($I - J$)	Std. error	Sig.	Lower bound	Upper bound
1.00	0.00	0.21667	0.25953	0.819	−0.4599	0.8932
2.00	0.00	0.95000*	0.25953	0.004	0.2734	1.6266
4.00	0.00	1.56667*	0.25953	0.000	0.8901	2.2432
8.00	0.00	0.30000	0.25953	0.608	−0.3766	0.9766

*The mean difference is significant at the 0.05 level.
[a]Dunnett t tests treat one group as a control, and compare all other groups against it.

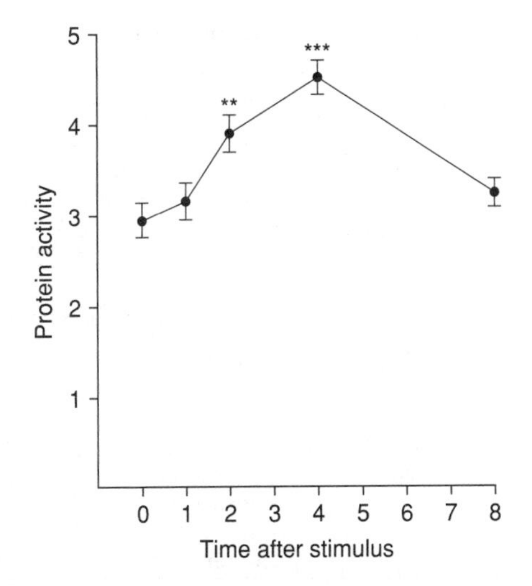

Figure A2 Graph showing the mean ± standard error of calcium-binding protein activity before and at various times after being given a heat stimulus (for each group $n = 6$). At 2 and 4 hours after heat shock activity was significantly higher than before: * $P < 0.05$; ** $P < 0.01$; *** $P < 0.001$.

Looking at the Sig. column, shows that there is a significant difference from the control (0 hours) after 2 and 4 hours (Sig. = 0.004 and 0.000 respectively) but not at 1 or 8 hours (Sig. = 0.819 and 0.608 respectively).

Therefore activity is significantly increased but only after more than 1 hour, and it drops off again before 8 hours.

Problem 5.2

Looking at the degrees of freedom (DF), it is clear that $4 + 1 = 5$ groups must have been examined, and in total $29 + 1 = 30$ observations must have been made. F is quite small (1.71) and Sig. is high ($0.35 > 0.05$). Therefore there was no significant difference between the groups.

Problem 5.3

The null hypothesis is that the mean aluminium concentration is the same at each point in time. To test whether this is the case you should carry out a repeated measures ANOVA in SPSS, as well as determine the descriptive statistics for the five time points.

SPSS will produce the following results (among other stuff).

Descriptive statistics

	Mean	Std. deviation	*N*
Week 1	14.7000	1.03785	8
Week 2	11.5625	0.97678	8
Week 3	9.6375	0.95609	8
Week 4	8.4875	1.19575	8
Week 5	8.0000	1.26604	8

Tests of within-subjects effects

Measure: MEASURE_1

Source		Type III sum of squares	df	Mean square	*F*	Sig.
Week	Sphericity assumed	238.484	4	59.621	157.482	0.000
	Greenhouse–Geisser	238.484	2.218	107.527	157.482	0.000
	Huynh–Feldt	238.484	3.292	72.441	157.482	0.000
	Lower-bound	238.484	1.000	238.484	157.482	0.000
Error(week)	Sphericity assumed	10.601	28	0.379		
	Greenhouse–Geisser	10.601	15.525	0.683		
	Huynh–Feldt	10.601	23.045	0.460		
	Lower-bound	10.601	7.000	1.514		

The descriptives table shows you that the mean aluminium activity is falling throughout the time period, from 14.7 at week 1 to 8.0 at week 5, but are these changes significant? Looking at the tests of within-subjects effects, table, for **Sphericity Assumed**

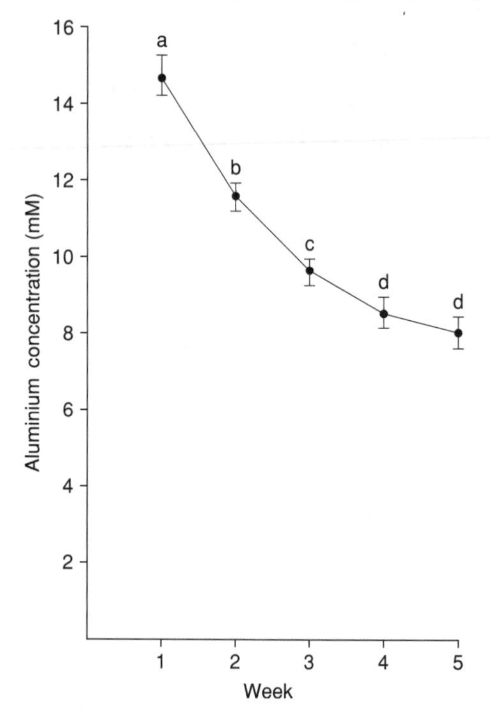

Figure A3 Graph showing the aluminium concentration in tanks at 5 weekly intervals after 20 snails had been placed in them (n = 8). Letters denote significant differences between groups. Groups denoted by the same letter are not significantly different from each other.

$F = 157.5$ and Sig. = 0.000 0.000, well below the 0.05 level needed for significance. Therefore aluminium levels are different at different times. But to test whether activity continues to fall throughout the experiment we must perform a **Bonferroni** test, which compares all time points with all the others. This comes up with the following results.

Pairwise comparisons

Measure: MEASURE_1

(*I*) week	(*J*) week	Mean difference (*I* – *J*)	Std. error	Sig.[a]	95% Confidence interval for difference[a]	
					Lower bound	Upper bound
1	2	3.138*	0.395	0.001	1.544	4.731
	3	5.063*	0.389	0.000	3.495	6.630
	4	6.213*	0.347	0.000	4.814	7.611
	5	6.700*	0.273	0.000	5.602	7.798
2	1	−3.138*	0.395	0.001	−4.731	−1.544
	3	1.925*	0.206	0.000	1.095	2.755
	4	3.075*	0.306	0.000	1.843	4.307
	5	3.563*	0.380	0.000	2.030	5.095
3	1	−5.063*	0.389	0.000	−6.630	−3.495
	2	−1.925*	0.206	0.000	−2.755	−1.095
	4	1.150*	0.211	0.010	0.299	2.001
	5	1.638*	0.306	0.011	0.403	2.872
4	1	−6.213*	0.347	0.000	−7.611	−4.814
	2	−3.075*	0.306	0.000	−4.307	−1.843
	3	−1.150*	0.211	0.010	−2.001	−0.299
	5	0.488	0.157	0.174	−0.147	1.122
5	1	−6.700*	0.273	0.000	−7.798	−5.602
	2	−3.563*	0.380	0.000	−5.095	−2.030
	3	−1.638*	0.306	0.011	−2.872	−0.403
	4	−0.488	0.157	0.174	−1.122	0.147

Based on estimated marginal means.
*The mean difference is significant at the 0.05 level.
[a]Adjustment for multiple comparisons: Bonferroni.

Here it can be seen that all time points are significantly different from each other (Sig. < 0.05) except for weeks 4 and 5 where Sig. = 0.174 which is greater than 0.05.

Therefore the aluminium levels continue to fall significantly only up to week 4 after which they level off.

Problem 5.4

The null hypothesis is that there is no difference between the numbers of colonies on dishes smeared with different antibiotics. Carrying out the Kruskall–Wallis test in SPSS gives the following results.

Kruskal-Wallis Test

Ranks

	Antibiotic	*N*	Mean rank
Number	0.00	10	25.80
	1.00	10	15.10
	2.00	10	21.60
	3.00	10	19.50
	Total	40	

Test statistics[a,b]

	Number
Chi-square	4.604
df	3
Asymp. sig.	0.203

[a]Kruskal–wallis test.
[b]Grouping variable: antibiotic.

The first table shows that the mean ranks of the four treatments are certainly different, but are these differences significant? The second table gives the value of chi-square as 4.604, which is well below the value of 7.815 need for 3 degrees of freedom (see Table S3). SPSS and MINITAB also directly calculate the significance probability as 0.203, which is well above the value of 0.05 needed for significance.

Therefore there is no significant difference between the numbers of colonies growing on plates which have had the different antibiotic treatments.

Problem 5.5

The null hypothesis is that there was no difference in the mood of the patients at the four times before and after taking the drug. Carrying out a Friedman's test using SPSS gives the following results.

Descriptive statistics

		Percentiles		
	N	25th	50th (Median)	75th
Before	10	1.7500	2.0000	3.0000
1 day	10	3.0000	3.5000	4.0000
1 week	10	3.0000	3.0000	4.0000
1 month	10	2.0000	2.0000	3.0000

Ranks

	Mean rank
Before	1.65
1 day	3.35
1 week	3.25
1 month	1.75

Test statistics[a]

N	10
Chi-square	18.805
df	3
Asymp. sig.	0.000

[a]Friedman test.

The descriptive statistics suggest that the mood of the patients improved 1 day and 1 week after taking the drug, but were these differences significant? The second table gives the mean ranks, and the third table gives the value of chi-square (18.805) which is well above the value of 7.800 need for four groups and ten blocks (see Table S6). SPSS and MINITAB also directly calculate the significance probability as 0.000, which is well below the value of 0.05 needed for significance.

Therefore the mood of the patients is significantly different at different times, and it looks as if it improves mood for over a week.

Problem 5.6

(a) To investigate which effects are significant we must carry out a two-way ANOVA. In SPSS this comes up with the following results.

Tests of between-subjects effects

Dependent variable: yield

Source	Type III sum of squares	df	Mean square	*F*	Sig.
Corrected model	336.246[a]	5	67.249	116.486	0.000
Intercept	2429.423	1	2429.423	4208.142	0.000
variety	0.145	1	0.145	0.251	0.618
nitrogen	240.911	2	120.456	208.648	0.000
variety * nitrogen	95.189	2	47.595	82.441	0.000
Error	27.711	48	0.577		
Total	2793.380	54			
Corrected total	363.957	53			

[a]R squared = 0.924 (adjusted r squared = 0.916).

It can be seen from this table that nitrogen has a significant effect on yield (Sig. = 0.000) as does the interaction between nitrogen and variety (Sig. = 0.000). However, variety does not have a significant effect (Sig. = 0.618 which is greater than 0.05). What does this mean? Well, we can find out by looking at the descriptive statistics and a plot of the means (Figure A4) for each variety at each nitrogen level (below).

Descriptive statistics

Dependent variable: yield

Variety	Nitrogen	Mean	Std. deviation	*N*
1.00	1.00	5.4667	0.57879	9
	2.00	7.1667	0.87178	9
	3.00	7.3333	0.66895	9
	Total	6.6556	1.10151	27
2.00	1.00	2.3889	0.62539	9
	2.00	7.1333	0.94868	9
	3.00	10.7556	0.79390	9
	Total	6.7593	3.57483	27
Total	1.00	3.9278	1.68795	18
	2.00	7.1500	0.88401	18
	3.00	9.0444	1.89929	18
	Total	6.7074	2.62052	54

Looking at the table and the plot, it is clear that adding nitrogen increases yield in both varieties. This is the cause of the significant term for nitrogen level. The average

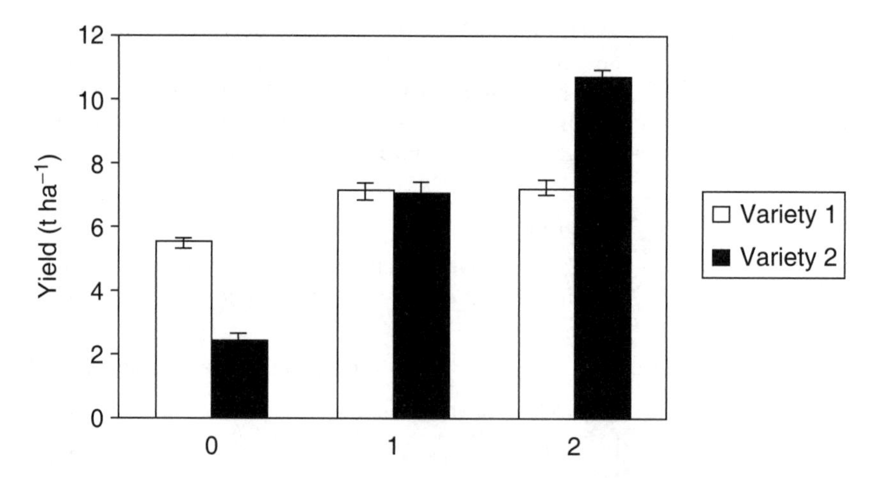

Figure A4 Mean ± standard error of yields for two different varieties of wheat at applications of nitrate of 0, 1 and 2 (kg m^{-2}).

yield of the two varieties is about the same (the cause of the non-significant variety term), but nitrogen has more effect on yield in Hereward than in Widgeon (the cause of the significant interaction term). Hence Widgeon does better without nitrogen; Hereward does better with lots of nitrogen.

Chapter 6

Problem 6.1

(a) Plot cell number against time.

(b) Plot chick pecking order against parent pecking order, since the behaviour and health of chicks is more likely to be affected by their parents than vice versa.

(c) Plot weight against height, because weight is more likely to be affected by height than vice versa.

(d) You can plot this graph either way, because length and breadth are a measure of size and may both be affected by the same factors.

Problem 6.2

The null hypothesis is that there is no linear association between leaf area and stomatal density. In the correlation analysis, SPSS and MINITAB both calculate a correlation coefficient r of -0.944. This looks like a strong negative correlation, but is it significant? SPSS and MINITAB give the significance probability as 0.000, well below the 0.05 needed for significance.

Therefore, leaf area and stomatal density show a significant negative correlation.

Problem 6.3

(a) Bone density is the dependent variable, so it should be plotted along the vertical axis. The SPSS graph is given below.

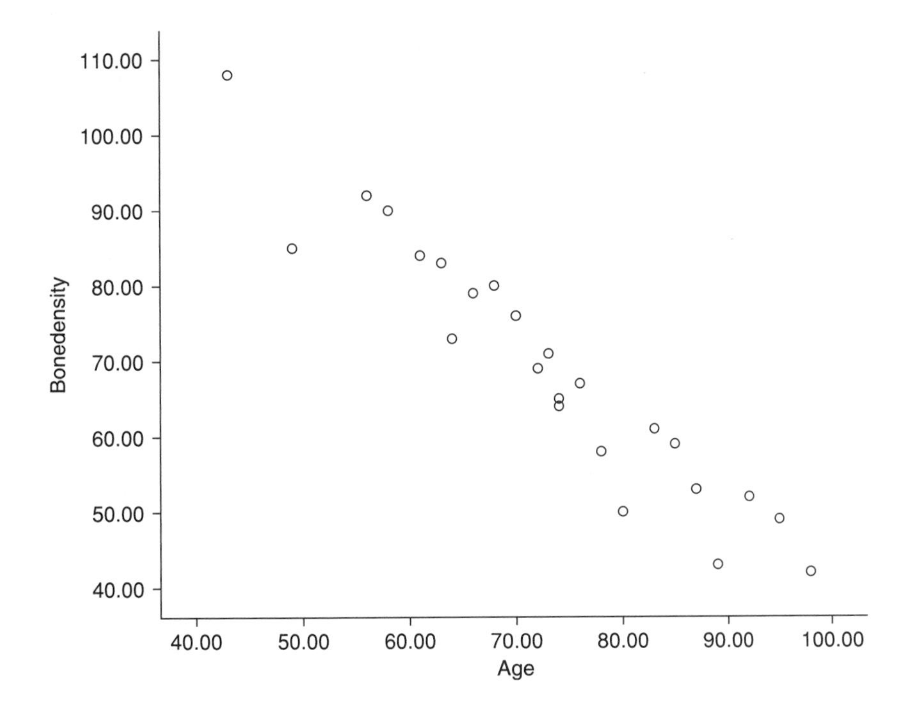

(b) SPSS comes up with the following results for the regression analysis amongst other tables.

Coefficients[a]

Model		Unstandardized coefficients		Standardized coefficients		
		B	Std. error	Beta	*t*	Sig.
1	(Constant)	151.277	5.637		26.835	0.000
	age	−1.128	0.076	−0.954	−14.880	0.000

[a]Dependent variable: bonedensity.

And MINITAB simply comes up with the regression equation

$$\text{Bone density} = 151.227 - (1.128 \times \text{Age})$$

From the graph and the above equation it appears that bone density falls significantly with age. To determine whether the fall is significant, we must examine the results of the t test for age. Here $t = -14.880$ and the significance probability = 0.000, well below the value of 0.05 needed for significance.

Therefore the slope is significantly different from 0. We can say there is a significant fall in bone density with age.

(c) Expected density at age 70 is found by inserting the value of 70 into the regression equation:

$$\text{Density} = 151.277 - (1.127\,51 \times 70) = 72.3$$

Problem 6.4

(a) Worm zinc concentration depends on environmental zinc concentration, not vice versa, so worm zinc concentration must be plotted along the y-axis. SPSS plots the following graph.

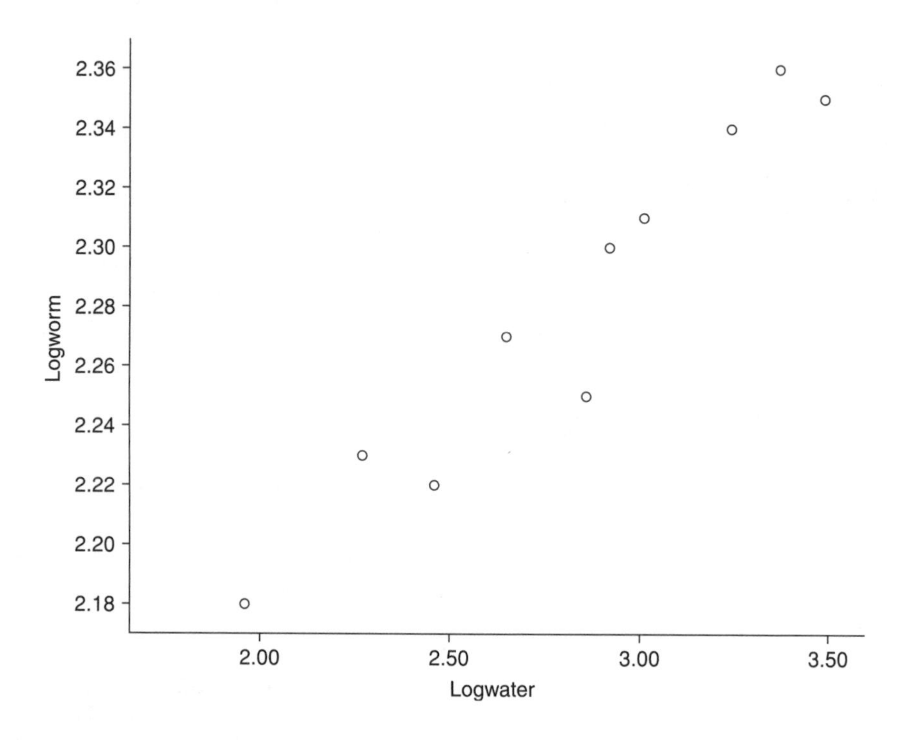

(b) In SPSS the regression analysis comes up with the following table amongst others.

Coefficients[a]

Model		Unstandardized coefficients		Standardized coefficients		
		B	Std. error	Beta	*t*	Sig.
1	(Constant)	1.945	0.033		58.191	0.000
	logwater	0.119	0.012	0.964	10.206	0.000

[a]Dependent variable: logworm.

And MINITAB simply comes up with the regression equation

$$\log[\mathrm{Zn}]_{\mathrm{worm}} = 1.1945 + (0.119 \times \log[\mathrm{Zn}]_{\mathrm{water}})$$

It is clear that the zinc concentration in the worms does increase with the zinc concentration in the water, but the slope is much lower than 1, being only 0.119. To investigate whether the slope is significantly different from 1 we must test the null hypothesis that the actual slope equals 1. To do this we carry out the following t test:

$$t = \frac{\text{Actual slope} - \text{Expected slope}}{\text{Standard deviation of slope}}$$

Here $t = (0.119 - 1)/0.012 = -73.4$. Its absolute value, 73.4, is much greater than the value of 2.306 needed for a significant effect at $10 - 2 = 8$ degrees of freedom. Therefore the slope is significantly different from 1 (less in fact). It is clear that the worms must be actively controlling their internal zinc concentrations.

Problem 6.5

(a) The relationship between seeding rate and yield as drawn by SPSS is shown below. It looks as if the yield rises to a maximum at a seeding rate of 300 m^{-2}, before falling again.

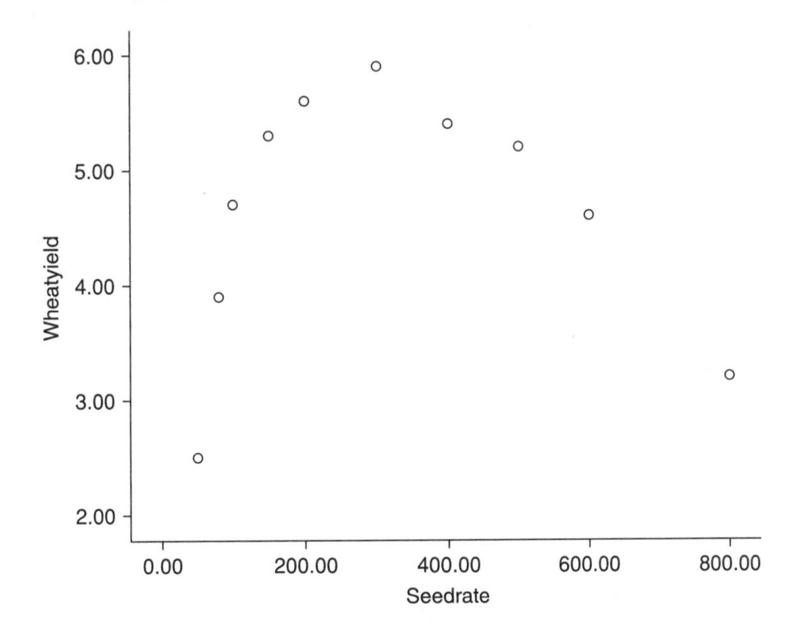

(b) In SPSS the regression analysis yields the following results among others:

Coefficients[a]

Model		Unstandardized coefficients		Standardized coefficients		
		B	Std. error	Beta	*t*	Sig.
1	(Constant)	4.659	0.619		7.531	0.000
	seedrate	–9.0E-005	0.002	–0.020	–0.058	0.955

[a]Dependent variable: wheatyield.

And MINITAB simply comes up with the regression equation

$$\text{Wheatyield} = 4.659 - (0.00009 \times \text{Seedrate})$$

It is clear that the slope is not significantly different from 0 (Sig. = 0.955 which is much greater than 0.05) and in fact the regression equation explains essentially none of the variability ($r^2 = 0.000$).

(c) There is no significant linear relationship between seeding rate and yield; the relationship is curvilinear. The moral of this exercise is that linear relationships are not the only ones you can get, so it is important to examine the data graphically.

Problem 6.6

(a) If $\log_{10} A = 0.3 + 2.36 \log_{10} L$, taking inverse logarithms gives

$$A = 10^{0.3} \times L^{2.36}$$
$$A = 2.0L^{2.36}$$

(b) If $\log_e N = 2.3 + 0.1T$, taking inverse natural logarithms gives

$$N = e^{2.3} \times e^{0.1T}$$
$$N = 10e^{0.1T}$$

Problem 6.7

(a) The logged data are shown in an SPSS graph below. lnmetabolism is plotted against temperature because metabolism can be affected by temperature and not vice versa.

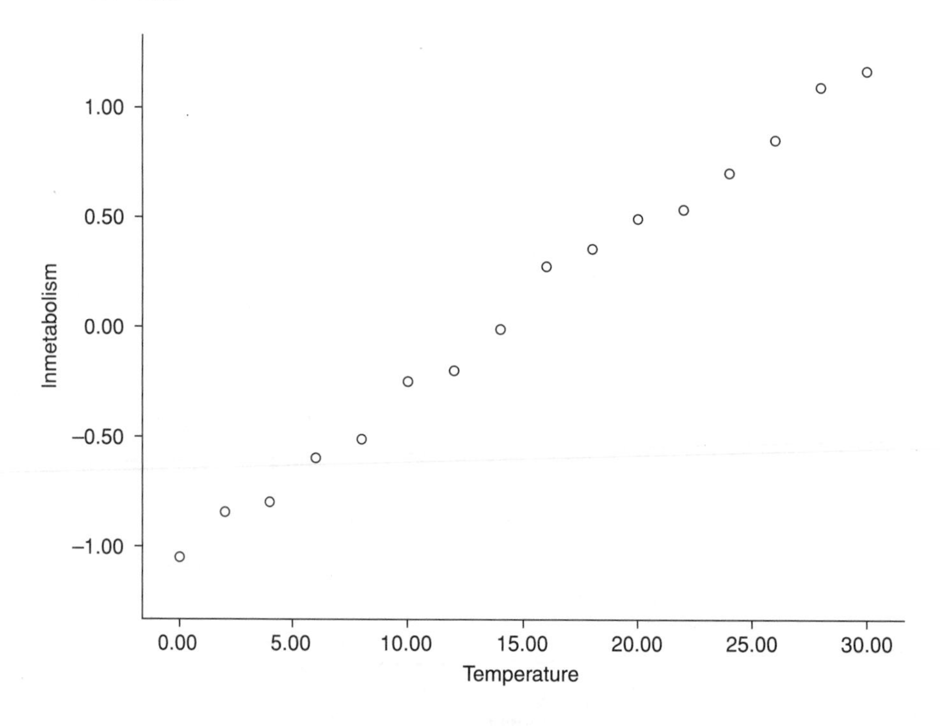

(b) Regression analysis in SPSS gives the following results.

Coefficients[a]

Model		Unstandardized coefficients		Standardized coefficients		
		B	Std. error	Beta	*t*	Sig.
1	(Constant)	–1.041	0.028		–37.692	0.000
	temperature	0.075	0.002	0.997	47.511	0.000

[a]Dependent variable: lnmetabolism.

And MINITAB simply comes up with the regression equation

$$\text{lnmetabolism} = -1.041 + (0.075 \times \text{Temperature})$$

The rise in metabolic rate with temperature is clearly significant because $t = 47.511$ and Sig. = 0.000 which is well below 0.05.

Converting back to real numbers, metabolism $= 0.30_e^{0.075 \times \text{temperature}}$.

Problem 6.8

The null hypothesis is that there was no correlation between dominance rank and testosterone level. Carrying out a Spearman's rank correlation in SPSS gives the value of ρ as −0.375. However, 0.375 is less than the critical value for 18 degrees of freedom of 0.472. SPSS and MINITAB also directly calculate the significance probability as 0.104, well above the level of 0.05 needed for significance.

Therefore there is no significant correlation between the dominance rank of males and their testosterone levels.

Chapter 7

Problem 7.1

The null hypothesis is that the mice are equally likely to turn to the right as to the left. Therefore the expected ratio is 1:1.

(a) After 10 trials, expected values are 5 towards the scent and 5 away. $\chi^2 = (-2)^2/5 + 2^2/5 = 0.80 + 0.80 = 1.60$ to two decimal places. This is below the critical value of 3.84 needed for significance for 1 degree of freedom, so there is as yet no evidence of a reaction.

(b) After 100 trials, expected values are 50 towards the scent and 50 away. $\chi^2 = (-16)^2/50 + 16^2/50 = 5.12 + 5.12 = 10.24$. This is greater than the critical value of 3.84 needed for significance for 1 degree of freedom, so there is clear evidence of a reaction. The mice seem to avoid the scent, in fact.

This problem shows the importance of taking large samples, as this will improve the chances of detecting effects.

Problem 7.2

The null hypothesis is that there is no linkage, so in this sample of 160 plants the expected numbers in each class are 90, 30, 30 and 10. Therefore $\chi^2 = (-3)^2/90 + 4^2/30 + (-2)^2/30 + 1^2/10 = 0.100 + 0.533 + 0.133 + 0.100 = 0.87$ to two decimal places. This is below the critical value of 7.34 needed for significance for 3 degrees of freedom, so there is no evidence of a ratio different from 9:3:3:1 and so no evidence of linkage.

Problem 7.3

The null hypothesis is that the incidence of illness in the town was the same as for the whole country. The expected values for illness in the town = 3.5% of 165 = 5.8, and therefore 159.2 without the illness. $\chi^2 = (9 - 5.8)^2/5.8 + (156 - 159.2)^2/159.2 = 3.2^2/5.8 + (-3.2)^2/159.2 = 1.76 + 0.06 = 1.82$. This is below the critical value of 3.84 needed for significance for 1 degree of freedom, so there is no evidence of a different rate of illness.

Problem 7.4

The null hypothesis is that the insects are randomly distributed about different coloured flowers. The completed table with expected values is given here, and it shows that in many cases the numbers of insects on flowers of particular colours is very different from expected. But is this a significant association? A χ^2 test is needed.

Insect visitors	Flower colour			Total
	White	Yellow	Blue	
Beetles	56 (26.04)	34 (38.16)	12 (37.80)	102
Flies	31 (32.43)	74 (47.51)	22 (47.06)	127
Bees and wasps	57 (85.53)	103 (125.33)	175 (124.14)	335
Total	144	211	209	564

(a) Working with four decimal places for the expected values, e.g. 26.0426 instead of 26.04, we obtain

$$\begin{aligned}\chi^2 &= 34.461 + 0.453 + 17.608 + 0.063 + \\ &= 14.767 + 13.346 + 9.518 + 3.978 + 20.837 \\ &= 115.031 = 115.03 \quad \text{(two decimal places)}\end{aligned}$$

This value is higher than the critical value of 9.48 needed for $(3 - 1) \times (3 - 1) = 4$ degrees of freedom. SPSS and MINITAB also directly calculate a significance probability of 0.000. Therefore we can conclude there is a significant association between insect type and flower colour.

(b) The highest χ^2 values are 34.46 for beetles and white flowers; 20.84 for bees and wasps and blue flowers; and 17.61 for beetles and blue flowers. Looking at the values, more beetles are found at white flowers than expected, so beetles in particular favour them; similarly, bees and wasps favour blue flowers; but beetles seem to avoid blue flowers.

Problem 7.5

The null hypothesis is that people with and without freckles have the same incidence of cancer. The completed contingency table is as shown and seems to indicate that the number of people with freckles who have cancer is greater than expected.

	Healthy	Cancer	Total
Freckles	945 (957.3)	33 (20.7)	978
No freckles	4981 (4968.7)	95 (107.3)	5076
Total	5926	128	6054

But is this a significant effect? A χ^2 test gives the following result: working with four decimal places for the expected values, we obtain

$$\begin{aligned}\chi^2 &= 0.159 + 7.343 + 0.031 + 1.415 = 8.948 \\ &= 8.95 \quad \text{(two decimal places)}\end{aligned}$$

This value is higher than the critical value of 3.84 needed for 1 degree of freedom. SPSS and MINITAB also calculate a significance probability of 0.003. Therefore there is a significant association between possession of freckles and skin cancer. More people with freckles get the disease.

Problem 7.6

The first thing to do is to calculate the number of ponds with neither newt species. This equals 745 – (180 + 56 + 236) = 273. The null hypothesis is that there is no association between the presence in ponds of smooth and palmate newts. The completed contingency table is shown here and appears to indicate that there are far more ponds than expected with both species or with neither species present.

	– Smooth	+ Smooth	Total
–Palmate	273 (200.05)	180 (252.95)	453
+Palmate	56 (128.95)	236 (163.05)	292
Total	329	416	745

Working with expected values to four decimal places, a χ^2 test gives the following results:

$$\chi^2 = 26.602 + 21.039 + 41.270 + 32.639 = 121.550$$
$$= 121.55 \quad \text{(two decimal places)}$$

The value is higher than the critical value of 3.84 needed for $(2-1) \times (2-1) = 1$ degree of freedom. SPSS and MINITAB also calculate a significance probability of 0.000. Therefore there is a significant association between the presence of the two species. In fact, the newts seem to be positively associated with each other. When one species is present, it is more likely that the other species will be present as well.

Chapter 8

Problem 8.1

Her null hypothesis is that there is no significant relationship between energy intake and heart rate. The statistical test to use is correlation. She is looking at measurements, looking for an association between two sets of measurements, and neither variable is clearly independent of the other.

Problem 8.2

His null hypothesis is that there is no significant association between particular habitats and species. The statistical test to use is the χ^2 test for association. He is looking at frequencies in different categories, and looking for an association between two types of category (species and habitat).

Problem 8.3

His null hypothesis is that there is no significant difference between the insulin levels of the three races. The statistical test to use is one-way ANOVA. He has taken measurements and is looking for differences between groups; there are more than two groups; the measurements are not matched; and he is investigating just one factor (race).

Problem 8.4

Her null hypothesis is that there is no significant difference between the expected ratio of numbers of the groups of snails and the expected 9:3:3:1 ratio. The statistical test to use is the χ^2 test for differences. She is dealing with frequencies in different categories, and there are expected frequencies (9:3:3:1).

Problem 8.5

His null hypothesis is that there is no significant difference in pesticide levels between the birds at different times of year. The statistical test to use is repeated measures ANOVA. He has taken measurements and is looking for differences between different sets of measurements; there are more than two sets of measurements and these are related, since they are taken on the same birds. Finally pesticide levels are continuous variables which are likely to be normally distributed.

Problem 8.6

Their null hypothesis is that there is no significant difference between blood pressure before and after taking the drug. The statistical test to use is the paired t test. They are looking at measurements and looking for differences; they will compare only two groups (before and after), and measurements will be in matched pairs (before and after).

Problem 8.7

The net production of oxygen by plants via photosynthesis results in them growing. Therefore if we can estimate how much a pot plant grows, we can estimate how much oxygen it produces. Let's suppose it grows at the (fast) rate of 1 g dry mass per day (so after a year it would have a wet weight of over a kilogram).

Now oxygen is produced by the following reaction:

$$6CO_2 + 6H_2O \rightarrow C_6H_{12}O_6 + 6O_2$$

But 1 mol of glucose weighs $(12 \times 6) + 12 + (16 \times 6) = 180$ g, so the number of moles of glucose produced per day $= 1/180 = 5.556 \times 10^{-3}$. For every 1 mol of dry matter produced, 6 mol of O_2 is also produced. Therefore for every 1 g of dry mass produced by the plant, the number of moles of oxygen produced $= 6/180 = 3.333 \times 10^{-2}$.

Since 1 mol of oxygen takes up 24 l, this makes up a volume of $3.333 \times 10^{-2} \times 24 = 0.80\ \text{l} = 0.8 \times 10^{-3}\ \text{m}^3 = 8 \times 10^{-4}\ \text{m}^3$. How does this compare with the amount of oxygen in the room. Well, let's imagine a room of $5\ \text{m} \times 4\ \text{m} \times 2.5$ m high, containing 20% oxygen. The volume of oxygen $= 5 \times 4 \times 2.5 \times 0.2 = 10\ \text{m}^3$. This is over 10 000 times greater. The tiny contribution of the plant will be far too small to make a difference. There is no point in doing the experiment.

Problem 8.8

The number of replicates required $\approx 9 \times (7.9/5)^2 \approx 22$. MINITAB also gives a figure of 22.

Problem 8.9

Number $\approx 16 \times (0.36/0.25)^2 \approx 33$. MINITAB gives a figure of 34.

Problem 8.10

A doubling of the risk means an increase of 0.035. Therefore $N \approx (9 \times 0.035 \times 0.965)/0.035^2 + 1 \approx 249$.

Problem 8.11

The first thing to do is to arrange for replication in your experiment. Each treatment should be given four plots. Next you must decide how to arrange the treatments around the plots. You could randomise totally, arranging treatments randomly in each of the 16 plots. However, in this case one treatment might tend to be restricted to one end of the site. A better solution is to split the site into four 2 m × 2 m blocks and randomise each of the four treatments within each block (see diagram below).

0	3.5	7	14	3.5	14	0	7
7	14	3.5	0	0	7	3.5	14

Next you must calculate how much fertiliser to apply to each plot. A litre of 1 *M* ammonium nitrate will contain 1 mol of the substance. The formula of ammonium nitrate is NH_4NO_3, so this will contain 2 mol of nitrogen, a mass of 28 g (the relative atomic mass of nitrogen is 14). Therefore the mass of nitrogen in 1 l of 20×10^{-3} *M* ammonium nitrate fertiliser is given by

$$\text{Mass } N = 1 \times 0.020 \times 28 = 0.56 \text{ g}$$

To supply 14 g the volume required = 14/0.56 = 25 l.

Each plot has area 1 m^2 so must be supplied with 25 × 1 = 25 l per year. But this must be spread over 25 applications, so the volume which must be applied each visit is 25/25 = 1 l of fertiliser.

What about the other plots? You could apply 0.5, 0.25 and 0 l of fertiliser to get the correct rates of 7, 3.5 and 0 g of nitrogen per year. However, you would be adding different quantities of water to each plot! To control this, you should add 1 l of fertiliser diluted by a factor of 2 and 4 respectively to the 7 and 3.5 g plots, and 1 l of water to each zero nitrogen plot.

Chapter 10

Problem 10.1

(a) m^2
(b) m s^{-1} (though the number will obviously be very low!)
(c) m^{-3} (number per unit volume)
(d) no units (it's one concentration divided by another)

Problem 10.2

(a) 192 MN or 0.192 GN
(b) 102 μg or 0.102 mg
(c) 0.12 ms (120 μs would imply that you had measured to three significant figures)
(d) 213 mm or 0.213 m

Problem 10.3

(a) 4.61×10^{-5} J
(b) 4.61×10^{8} s

Problem 10.4

(a) 3.81×10^{9} Pa
(b) 4.53×10^{-3} W

(c) 3.64×10^{-1} J
(d) 4.8×10^{-6} kg (remember that the SI unit of mass is the kg)
(e) 2.1×10^{-16} kg (remember that the SI unit of mass is the kg)

Problem 10.5

(a) $250 \times 10^3 = 2.50 \times 10^5$ kg
(b) $0.3 \times 10^5 = 3 \times 10^4$ Pa
(c) $24 \times 10^{-10} = 2.4 \times 10^{-9}$ m

Problem 10.6

In each case use a similar degree of precision as the original measurements.

(a) $35 \times 0.9144 = 32.004$ m
$= 32$ m (two significant figures)
(b) $(3 \times 0.3048) + (3 \times 2.54 \times 10^{-2}) = 0.9144 + 0.0762$
$= 0.99$ m (two significant figures)
(c) $9.5 \times (0.9144)^2 = 7.943$ m^2
$= 7.9$ m^2 (two significant figures)

Problem 10.7

(a) $(1.23 \times 2.456) \times 10^{(3+5)}$ m$^2 = 3.02 \times 10^8$ m^2 (three significant figures)
(b) $(2.1/4.5) \times 10^{(-2+4)}$ J kg$^{-1} = 0.4666 \times 10^2$
$= 4.7 \times 10^1$ J kg^{-1} (two significant figures)

Problem 10.8

(a) 1.3 mmol
(b) 365 MJ or 0.365 GJ
(c) 0.24 μm (not 240 nm, because this implies knowledge to three significant figures)

Problem 10.9

The concentration is the number of cells divided by the volume in which they were found. The dimensions of the box are 1×10^{-3} m by 1×10^{-3} m by 1×10^{-4} m. Therefore its volume is $1 \times 10^{(-3-3-4)} = 1 \times 10^{-10}$ m^3. The concentration of blood cells is therefore

$$
\begin{aligned}
652/(1 \times 10^{-10}) &= (6.52 \times 10^2)/(1 \times 10^{-10}) \\
&= (6.52/1) \times 10^{(2+10)} \\
&= 6.52 \times 10^{12} \text{ m}^{-3}
\end{aligned}
$$

Problem 10.10

The volume of water which had fallen is the depth of water which had fallen multiplied by the area over which it had fallen.

$$
\begin{aligned}
\text{Depth} &= 0.6 \times 2.54 \times 10^{-2} \\
&= 1.524 \times 10^{-2} \text{ m} \\
\text{Area} &= 2.6 \times 10^4 \text{ m}^2 \\
\text{Therefore volume} &= 1.524 \times 10^{-2} \times 2.6 \times 10^4 \\
&= 3.962 \times 10^2 \text{ m}^3 \\
&= 4 \times 10^2 \text{ m}^3 \quad \text{(one significant figure)}
\end{aligned}
$$

Problem 10.11

The concentration is the number of moles of glucose per litre.

$$\begin{aligned}\text{Number of moles} &= \frac{\text{Mass in grams}}{\text{Molecular mass}}\\ &= \frac{25}{(6 \times 12) + (12 \times 1) + (6 \times 16)}\\ &= 25/180 = 1.3888 \times 10^{-1}\\ \text{Concentration} &= 1.3888 \times 10^{-1}/2 = 6.9444 \times 10^{-2}\,M\\ &= 6.9 \times 10^{-2}\,M \quad \text{(two significant figures)}\end{aligned}$$

Problem 10.12

The first thing to work out is the volume of CO_2 that was produced.

$$\begin{aligned}\text{Volume } CO_2 \text{ produced} &= \text{Volume of air} \times \text{Proportion of it which is } CO_2\\ &= 45 \times 0.036 = 1.62\,\text{l}\end{aligned}$$

And we know that at room temperature and pressure 1 mol of gas takes up 24 l, so

$$\text{Number of moles } CO_2 = 1.62/24 = 6.75 \times 10^{-2}$$

The mass of CO_2 produced equals the number of moles multiplied by the mass of each mole of CO_2. Since the mass of 1 mol of $CO_2 = 12 + (2 \times 16) = 44$ g, we have

$$\begin{aligned}\text{Mass of gas produced} &= 6.75 \times 10^{-2} \times 44 = 2.97\text{ g}\\ &= 2.97 \times 10^{-3}\text{ kg}\end{aligned}$$

Production of this gas took 5 minutes $= 5 \times 60 = 300$ s, so

$$\begin{aligned}\text{Rate of gas production} &= (2.97 \times 10^{-3})/300\\ &= 9.9 \times 10^{-6}\text{ kg s}^{-1}\end{aligned}$$

Problem 10.13

The energy produced by the reaction was converted to heat, and heat energy = mass × specific heat × temperature rise. First, we need to work out the mass of water. Fortunately, this is easy as 1 l of water weighs 1 kg. Therefore 0.53 l has a mass of 0.53 kg. From Table 10.5 we can see that water has a specific heat of 4.2×10^3 J K^{-1} kg^{-1}, therefore

$$\begin{aligned}\text{Heat energy} &= 0.53 \times 4.2 \times 10^3 \times 2.4\text{ J}\\ &= 5342.4\text{ J}\\ &= 5.3 \times 10^3\text{ J or } 5.3\text{ kJ} \quad \text{(two significant figures)}\end{aligned}$$

Problem 10.14

The first thing to do is work out the number of moles of X you will use:

$$\begin{aligned}\text{Number of moles} &= \text{Volume (in litres)} \times \text{Concentration (in moles per litre)}\\ &= (8 \times 80 \times 10^{-3}) \times 3 \times 10^{-3}\\ &= 1.92 \times 10^{-3}\end{aligned}$$

Now obtain the mass of 1.92×10^{-3} mol of X:

$$\begin{aligned} \text{Mass (in grams)} &= \text{Number of moles} \times \text{Molecular mass} \\ &= 1.92 \times 10^{-3} \times 258 \\ &= 0.495 \text{ g} \end{aligned}$$

And finally the cost of 0.495 g of X:

$$\begin{aligned} \text{Cost} &= \text{Number of grams} \times \text{Price per gram} \\ &= 0.495 \times 56 \\ &= £28 \quad \text{(two significant figures)} \end{aligned}$$

Since this is far less than £1000 you will easily be able to afford it.

Problem 10.15

The first thing to do is to work out the volume of methane produced by bogs per year. We have

$$\text{Yearly production} = \text{Daily productivity} \times \text{Area of bog} \times \text{Days in a year}$$

And we have

$$\begin{aligned} \text{Daily productivity} &= 21 \text{ ml m}^{-2} = 2.1 \times 10^{-2} \text{ l m}^{-2} \\ \text{Area of bogs} &= 3.4 \times 10^{6} \text{ km}^2 = 3.4 \times 10^{12} \text{ m}^2 \end{aligned}$$

Therefore

$$\begin{aligned} \text{Yearly production} &= 2.1 \times 10^{-2} \times 365 \times 3.4 \times 10^{12} \\ &= 2.606 \times 10^{13} \text{ l} \end{aligned}$$

Next you need to work out how many moles this is equal to and hence the mass of methane produced per year. Since 1 mol of gas takes up 24 litres, we have

$$\begin{aligned} \text{Number of moles} &= \frac{\text{Volume (in litres)}}{24} \\ &= (2.606 \times 10^{13})/24 \\ &= 1.086 \times 10^{12} \end{aligned}$$

We also have

$$\text{Mass of methane (in grams)} = \text{Number of moles} \times \text{Molecular mass}$$

Since the molecular mass of methane is $12 + (1 \times 4) = 16$, therefore

$$\begin{aligned} \text{Mass of methane} &= 1.086 \times 10^{12} \times 16 \\ &= 1.737 \times 10^{13} \text{ g} \\ &= 1.737 \times 10^{10} \text{ kg} \end{aligned}$$

However, since this is three times as effective as CO_2, this is equivalent to $1.737 \times 10^{10} \times 3 = 5.2 \times 10^{10}$ kg of CO_2.

How does this compare with the amount of CO_2 produced by burning fossil fuels? This equals 25 Gt. We need to convert to kg:

$$25 \text{ Gt} = 25 \times 10^{9} \text{ t} = 25 \times 10^{12} \text{ kg} = 2.5 \times 10^{13} \text{ kg}$$

This is much more. The ratio of the effect of fossil fuel to the effect of bog methane production is

$$\frac{2.5 \times 10^{13}}{5.2 \times 10^{10}} = \sim 500$$

Therefore bog methane will have a negligible effect compared with our use of fossil fuels.

Problem 10.16

(a) 1.65
(b) 2.65
(c) −3.35
(d) 6
(e) 0

Problem 10.17

(a) 25.1
(b) 251
(c) 3.98×10^{-4}
(d) 10^4
(e) 1

Problem 10.18

(a) In 3×10^{-4} M HCl the concentration of H^+ is $[H^+] = 3 \times 10^{-4}$ M. Therefore pH = 3.5.
(b) In 4×10^{-6} M H_2SO_4 the concentration of H^+ is $[H^+] = 8 \times 10^{-6}$ M. Therefore pH = 5.1.

Problem 10.19

Concentration of H^+ ions in pH 2.1 = $10^{-2.1} = 7.94 \times 10^{-3}$ M. But each molecule of H_2SO_4 has two hydrogen ions. Therefore the concentration of $H_2SO_4 = (7.94 \times 10^{-3})/2 = 3.97 \times 10^{-3}$ M.

$$\begin{aligned} \text{Number of moles} &= \text{Concentration} \times \text{Volume} \\ &= 3.97 \times 10^{-3} \times 0.160 \\ &= 6.35 \times 10^{-4} \\ \text{Molecular mass of } H_2SO_4 &= 2 + 32 + 64 = 98 \\ \text{Mass of } H_2SO_4 &= \text{Number of moles} \times \text{Molecular mass (in grams)} \\ &= 6.35 \times 10^{-4} \times 98 \\ &= 6.22 \times 10^{-2}\ \text{g} \\ &= 6.2 \times 10^{-5}\ \text{kg} \quad \text{(two significant figures)} \end{aligned}$$

Problem 10.20

(a) 3.40
(b) −3.73
(c) 0

Problem 10.21

(a) 20.1
(b) 0.050
(c) 1

Statistical tables

Table S1 Critical values for the *t* statistic

Critical values of *t* at the 5%, 1% and 0.1% significance levels. Reject the null hypothesis if the absolute value of *t* is **greater than or equal to** the tabulated value at the chosen significance level, for the calculated number of degrees of freedom.

Degrees of freedom	Significance level		
	5%	1%	0.1%
1	12.706	63.657	636.619
2	4.303	9.925	31.598
3	3.182	5.841	12.941
4	2.776	4.604	8.610
5	2.571	4.032	6.859
6	2.447	3.707	5.959
7	2.365	3.499	5.405
8	2.306	3.355	5.041
9	2.262	3.250	4.781
10	2.228	3.169	4.587
11	2.201	3.106	4.437
12	2.179	3.055	4.318
13	2.160	3.012	4.221
14	2.145	2.977	4.140
15	2.131	2.947	4.073
16	2.120	2.921	4.015
17	2.110	2.898	3.965
18	2.101	2.878	3.922
19	2.093	2.861	3.883
20	2.086	2.845	3.850
21	2.080	2.831	3.819
22	2.074	2.819	3.792
23	2.069	2.807	3.767
24	2.064	2.797	3.745
25	2.060	2.787	3.725
26	2.056	2.779	3.707
27	2.052	2.771	3.690
28	2.048	2.763	3.674
29	2.045	2.756	3.659
30	2.042	2.750	3.646
40	2.021	2.704	3.551
60	2.000	2.660	3.460
120	1.980	2.617	3.373
∞	1.960	2.576	3.291

Table S2 Critical values for the correlation coefficient r

Critical values of the correlation coefficient r at the 5%, 1% and 0.1% significance levels. Reject the null hypothesis if your absolute value of r is **greater than or equal to** the tabulated value at the chosen significance level, for the calculated number of degrees of freedom.

Degrees of freedom	Significance level		
	5%	1%	0.1%
1	0.996 92	0.999 877	0.999 9988
2	0.950 00	0.990 000	0.999 00
3	0.8793	0.958 73	0.991 16
4	0.8114	0.917 20	0.974 06
5	0.7545	0.8745	0.950 74
6	0.7076	0.8343	0.924 93
7	0.6664	0.7977	0.8982
8	0.6319	0.7646	0.8721
9	0.6021	0.7348	0.8471
10	0.5760	0.7079	0.8233
11	0.5529	0.6835	0.8010
12	0.5324	0.6614	0.7800
13	0.5139	0.6411	0.7603
14	0.4973	0.6226	0.7420
15	0.4821	0.6055	0.6524
16	0.4683	0.5897	0.7084
17	0.4555	0.5751	0.6932
18	0.4438	0.5614	0.6787
19	0.4329	0.5487	0.6652
20	0.4427	0.5368	0.6524
25	0.3809	0.4869	0.6974
30	0.3494	0.4487	0.5541
35	0.3246	0.4182	0.5189
40	0.3044	0.3932	0.4896
45	0.2875	0.3721	0.4648
50	0.2732	0.3541	0.4433
60	0.2500	0.3248	0.4078
70	0.2319	0.3017	0.3799
80	0.2172	0.2830	0.3568
90	0.2050	0.2673	0.3375
100	0.1946	0.2540	0.3211

Table S3 Critical values for the χ^2 statistic

Critical values of χ^2 at the 5%, 1% and 0.1% significance levels. Reject the null hypothesis if your value of χ^2 is greater than or equal to than the tabulated value at the chosen significance level, for the calculated number of degrees of freedom.

Degrees of freedom	Significance level		
	5%	1%	0.1%
1	3.841	6.653	10.827
2	5.991	9.210	13.815
3	7.815	11.345	16.266
4	9.488	13.277	18.467
5	11.070	15.086	20.515
6	12.592	16.812	22.457
7	14.067	18.457	24.322
8	15.507	20.090	26.125
9	16.919	21.666	27.877
10	18.307	23.209	29.588
11	19.675	24.725	31.264
12	21.026	26.217	32.909
13	22.362	27.688	34.528
14	23.685	29.141	36.123
15	24.996	30.578	37.697
16	26.296	32.000	39.252
17	27.587	33.409	40.792
18	28.869	34.805	42.312
19	30.144	36.191	43.820
20	31.410	37.566	45.315
21	32.671	38.932	46.797
22	33.924	40.289	48.268
23	35.172	41.638	49.728
24	36.415	42.980	51.179
25	37.652	44.314	52.260
26	38.885	45.642	54.052
27	40.113	46.963	55.476
28	41.337	48.278	56.893
29	42.557	49.588	58.302
30	43.773	50.892	59.703

Table S4 Critical values for the Wilcoxon *T* distribution

Critical values of T at the 5%, 1% and 0.1% significance levels. Reject the null hypothesis if your value of T is **less than or equal to** than the tabulated value at the chosen significance level, for the calculated number of degrees of freedom.

Degrees of freedom	Significance level		
	5%	1%	0.1%
1			
2			
3			
4			
5			
6	0		
7	2		
8	3	0	
9	5	1	
10	8	3	
11	10	5	0
12	13	7	1
13	17	9	2
14	21	12	4
15	25	15	6
16	29	19	8
17	34	23	11
18	40	27	14
19	46	32	18
20	52	37	21
21	58	42	25
22	65	48	30
23	73	54	35
24	81	61	40
25	89	68	45
26	98	75	51
27	107	83	57
28	116	91	64
29	126	100	71
30	137	109	78
31	147	118	86
32	159	128	94
33	170	138	102
34	182	148	111
35	195	159	120
36	208	171	130
37	221	182	140

Degrees of freedom	Significance level		
	5%	1%	0.1%
38	235	194	150
39	249	207	161
40	264	220	172
41	279	233	183
42	294	247	195
43	310	261	207
44	327	276	220
45	343	291	233
46	361	307	246
47	378	322	260
48	396	339	274
49	415	355	289
50	434	373	304
51	453	390	319
52	473	408	335
53	494	427	351
54	514	445	368
55	536	465	385
56	557	484	402
57	579	504	420
58	602	525	438
59	625	546	457
60	648	567	476

Table S5 Critical values for the Mann–Whitney *U* distribution

Critical values of U at the 5% significance level. Reject the null hypothesis if your value of U is **less than or equal to** than the tabulated value, for the sizes of the two samples, u_1 and u_2.

n_2	n_1 1	2	3	4	5	6	7	8	9	10	11	12	13	14	15	16	17	18	19	20
1	–	–	–	–	–	–	–	–	–	–	–	–	–	–	–	–	–	–	–	–
2	–	–	–	–	–	–	–	0	0	0	0	1	1	1	1	1	2	2	2	2
3	–	–	–	0	0	1	1	2	2	3	3	4	4	5	5	6	6	7	7	8
4	–	–	0	0	1	2	3	4	4	5	6	7	8	9	10	11	11	12	13	14
5	–	–	0	1	2	3	5	6	7	8	9	11	12	13	14	15	17	18	19	20
6	–	–	1	2	3	5	6	8	10	11	13	14	16	17	19	21	22	24	25	27
7	–	–	1	3	5	6	8	10	12	14	16	18	20	22	24	26	28	30	32	34
8	–	0	2	4	6	8	10	13	15	17	19	22	24	26	29	31	34	36	38	41
9	–	0	2	4	7	10	12	15	17	20	23	26	28	31	34	37	39	42	45	48
10	–	0	3	5	8	11	14	17	20	23	26	29	33	36	39	42	45	48	52	55
11	–	0	3	6	9	13	16	19	23	26	30	33	37	40	44	47	51	55	58	62
12	–	1	4	7	11	14	18	22	26	29	33	37	41	45	49	53	57	61	65	69
13	–	1	4	8	12	16	20	24	28	33	37	41	45	50	54	59	63	67	72	76
14	–	1	5	9	13	17	22	26	31	36	40	45	50	55	59	64	67	74	78	83
15	–	1	5	10	14	19	24	29	34	39	44	49	54	59	64	70	75	80	85	90
16	–	1	6	11	15	21	26	31	37	42	47	53	59	64	70	75	81	86	92	98
17	–	2	6	11	17	22	28	34	39	45	51	57	63	67	75	81	87	93	99	105
18	–	2	7	12	18	24	30	36	42	48	55	61	67	74	80	86	93	99	106	112
19	–	2	7	13	19	25	32	38	45	52	58	65	72	78	85	92	99	106	113	119
20	–	2	8	14	20	27	34	41	48	55	62	69	76	83	90	98	105	112	119	127

Table S6 Critical values for the Friedman χ^2 distribution

Critical values of χ^2 at the 5%, 1% and 0.1% significance levels. Reject the null hypothesis if your value of χ^2 is **greater than or equal to** than the tabulated value, for a groups and b blocks.

Groups	Blocks	Significance level		
a	b	5%	1%	0.1%
3	2	–	–	–
3	3	6.000	–	–
3	4	6.500	8.000	–
3	5	6.400	8.400	10.000
3	6	7.000	9.000	12.000
3	7	7.143	8.857	12.286
3	8	6.250	9.000	12.250
3	9	6.222	9.556	12.667
3	10	6.200	9.600	12.600
3	11	6.545	9.455	13.273
3	12	6.167	9.500	12.500
3	13	6.000	9.385	12.923
3	14	6.143	9.000	13.286
3	15	6.400	8.933	12.933
4	2	6.000	–	–
4	3	7.400	9.000	–
4	4	7.800	9.600	11.100
4	5	7.800	9.960	12.600
4	6	7.600	10.200	12.800
4	7	7.800	10.371	13.800
4	8	7.650	10.350	13.800
4	9	7.800	10.867	14.467
4	10	7.800	10.800	14.640
4	11	7.909	11.073	14.891
4	12	7.900	11.100	15.000
4	13	7.985	11.123	15.277
4	14	7.886	11.143	15.257
4	15	8.040	11.240	15.400
5	2	7.600	8.000	–
5	3	8.533	10.133	11.467
5	4	8.800	11.200	13.200
5	5	8.960	11.680	14.400
5	6	9.067	11.867	15.200
5	7	9.143	12.114	15.657
5	8	9.300	12.300	16.000
5	9	9.244	12.444	16.356
5	10	9.280	12.480	16.480

Groups	Blocks	Significance level		
a	b	5%	1%	0.1%
6	2	9.143	9.714	–
6	3	9.857	11.762	13.286
6	4	10.286	12.714	15.286
6	5	10.486	13.229	16.429
6	6	10.571	13.619	17.048
6	7	10.674	13.857	17.612
6	8	10.714	14.000	18.000
6	9	10.778	14.143	18.270
6	10	10.800	14.299	18.514

Table S7 Critical values for the Spearman rank correlation coefficient ρ

Critical values of the correlation coefficient ρ at the 5%, 1% and 0.1% significance levels. Reject the null hypothesis if your absolute value of ρ is **greater than or equal to** the tabulated value at the chosen significance level, for the calculated number of degrees of freedom.

Degrees of freedom	Significance level		
	5%	1%	0.1%
1			
2			
3			
4			
5	1.000		
6	0.886	1.000	
7	0.786	0.929	1.000
8	0.738	0.881	0.976
9	0.700	0.833	0.933
10	0.648	0.794	0.903
11	0.618	0.755	0.873
12	0.587	0.727	0.846
13	0.560	0.703	0.824
14	0.538	0.679	0.802
15	0.521	0.654	0.779
16	0.503	0.635	0.762
17	0.485	0.615	0.748
18	0.472	0.600	0.728
19	0.460	0.584	0.712
20	0.447	0.570	0.696
25	0.398	0.511	0.630
30	0.362	0.467	0.580
35	0.335	0.433	0.539
40	0.313	0.405	0.507
45	0.294	0.382	0.479
50	0.279	0.363	0.456
60	0.255	0.331	0.418
70	0.235	0.307	0.388
80	0.220	0.287	0.363
90	0.207	0.271	0.343
100	0.197	0.257	0.326

Index

Note: definitions in the Glossary are indicated by **emboldened numbers**